Them That Live the Longest

More From the North-east Farm

Them That Live the Longest

More From the North-east Farm

CHARLIE ALLAN

BIRLINN

First published in 2010 by
Birlinn Limited
West Newington House
10 Newington Road
Edinburgh
EH9 1QS

www.birlinn.co.uk

ISBN 978 1 84158 890 2

British Library Cataloguing-in-Publication Data
A catalogue record for this book is available from the British Library

Typeset by Iolaire Typesetting, Newtonmore
Printed and bound by MPG Books Ltd, Bodmin

To my ancestors, in whose shadow I stand, bound by my duty to find another among us who will take Little Ardo, care for it, take increasing harvests from it and even come to love it, bound against the temptations offered by the fat lands across Ythanvale to the south.

Contents

CONTENTS

Little Ardo

This takes the family farm on the southern edge of Buchan from 1955, when a staff of six, five with farm cottages, were employed to produce milk, beef, potatoes, barley and wheat, up to the present, when it employs only the farmer, who rents out his cottages, and, with no regular help, produces beef and barley. He struggles on, producing more food with much more capital and far less profit . . . Or so he says.

The following illustrations by Rhoda Howie show how the farm has changed over the generations it has been owned by the family.

Little Ardo as the author's great-great grandfather William Yull found it when he arrived in 1837.

As it was when Maitland Mackie, the author's grandfather, took over in 1911, with the addition of William Yull's barn and John Yull's Victorian annexe to the house.

As it was when John R. Allan, the author's father, returned from the war in 1945. Note Maitland Mackie's extensive buildings – his new dairy byre, tattie shed, barn, piggery, and, the piece de resistance, the two wooden henhouses which almost came in at the back door.

What the author found when he took over in 1974. Note John Allan's great improvement had been scrapping Maitland Mackie's henhouses. The only building, apart from a low extension you cannot see to the house, being the new neepshed in the north-west corner of the midden.

What the author handed over to his daughter Sarah and her husband Neil Purdie in 1997. Note the fine Dutch barn to the north and the barley beef factory with its grain tower to the south-west and the open fronted shed in between. The old stable and the new neepshed had blown down.

How she stands as we go to press (2010), with enough buildings for a small airport. The author's son-in-law has put up six new sheds or bits of sheds and re-roofed most of the steading. Surely he has finished? And yet, there is talk of building new silage pits . . .

Preface

The title of the first volume of these memoirs, *The Truth Tells Twice,* came from my granny, Mrs Mary Mackie of North Ythsie, with whom I spent the last years of the Second World War. It was the saying with which she instilled truthfulness in her six children and nineteen grandchildren. The title of this volume comes from her husband, Maitland Mackie. He used, as the great changes he saw in his ninety years from 1886 to 1976 unfolded, to exclaim, 'them that live the langest see the maist ferlies' (wonders). He certainly saw many, from the arrival in Buchan of the first motor-car to sputniks in space and the day the minister's wife appeared in Tarves wearing a mini-skirt.

In *The Truth Tells Twice* I told you about my uncle, George Mackie, the war hero, who captained bombers. The worst trips were to Berlin, which was the most heavily defended target. He dreaded hearing that the next assignment was Berlin, but to help with morale he always whistled as he returned from the briefing to tell the crew the dreaded news, 'It's Berlin.' George was grounded after seventy-three sorties over enemy territory. They reasoned that it would be too great a blow to morale if one with his record were killed. His crew were all lost on the next sortie.

So the day came when, long after the war, Mackie had become a life peer and, as Liberal spokesman on agriculture, he went in a parliamentary delegation to see the Bundestag, the Brandenburg Gate, the Unter Den Linden avenue and the rest of the German capital which he had tried so hard to obliterate. One of his hosts, who had clearly neglected to do his homework, asked the now

Lord Mackie of Benshie, 'And so, Mister Mackie, haff you seen Berlin before?' In the wonderfully lugubrious RAF accent which he still does so well, George replied, 'Many times, but only from the air and at night.'

Had I known it, I would have included that story in *The Truth Tells Twice*. And there is much else which might have been included there, about which I was told only after people had read it – like the story of the maid a-courting in my grandparents' house at North Ythsie of Tarves. One of the decisions farmers and their wives had to make in the 1930s was whether to defend their hard-working live-in female staff from some of the very few pleasures available to them, those to be had with men. My grandmother was very keen to protect her girls' honour but Maitland, although sixty years an elder of a rather narrow-minded Church of Scotland, was a bit more relaxed about those moral problems. Anyway, according to their long-time neighbour Jack Sleigh of Tolquhon, my grand-father was awakened one night by an urgent shake from Mary, 'Maitland, Maitland. Wake up. There's a man in the maid's room.'

'What do you want me to do about it?' said her husband, tired after a hard day's work.

'Get up and stop them.'

'How do you know there's anything to stop?'

'I can hear them.'

'Well, if you can hear them, you're far too late,' said my grandfather, and went back to sleep.

And my Uncle Geordie himself told me a very surprising anecdote. It again throws a less than saintly light on the grandfather I idolise still, though his Good Shepherd has long since gathered him in. The maid who had the unenviable job of scrubbing the handsome red Peterhead granite step at North Ythsie, where I spent some of the war years in my grandparents' home, is said to have declared in a mixture of exasperation and disgust: 'This step smells like a bothy door. Maister Mackie and young Geordie should hudd away ower a bit fae the hoose.'

And then there's the letter I found among my parents' effects from Peter, one of those Italian prisoners of war who worked at

North Ythsie and was clearly smitten by the aura which grew up around my grandparents. Somebody had sent him a photo of me, and this is Peter's bread-and-butter letter:

> Dear Charles and family. I was comen back for my warek I saw the photo of Charles when I have been very glat to seen it again. It is long time that I didn't sent to your hall some of my news because a very misfortune have tack me very deep a month ago; having got my wife deat and she leave me with a little babe it is a female one. Exkus me of my Writen. I leave for the moment the next one will be longer. Ill sent to you all my best wishes for the year of 48 and gif my kind regards to your fader and mother, Mr & Mrs Mackie and the family Walker with all the North Ythsie friends. From your truly friend Peter Falletti.

What became of Peter I don't know, for as far as I know he never did write again. He clearly was more a friend than a prisoner of war. The 'Talies' all were.

Then I would have included, had I known it in time, the story of John Mackie, my great-grandfather and farmer at Mains of Elrick, when one day in 1880 he drew cash from the bank in Auchnagatt and was given £10 too much. There was a great stir of counting and checking in the bank when the deficiency was discovered. The next time John Mackie went into the bank, Fowler, the bank agent (as the managers of country branches were then known) asked if he had got £10 too much cash. John said, 'I really couldn't say. I just counted it up till I saw I had got what I asked for and I didna worry about the rest.'

A second chance to tell you those stories is part of my reason for writing another volume of memoirs. I must try not to make the same mistake again for there won't be a third.

CHAPTER ONE

Gentle Decline

1955–67

The Truth Tells Twice, the story of Little Ardo of which this is really volume two, finished in the mid '50s, with John Allan home from the war and firmly established as a gentleman farmer with James Low, one of the great grieves who were the making of Aberdeenshire. Low had quickly made an unflattering assessment of my father's usefulness as a farm worker and advised him to stick to the office, his writing and his broadcasting. The other lynchpin of the farm was James Kelman, his top-class dairyman. Those two were making a first-class job of the system which had fulfilled the farmer's first ambition for the place, the elimination of the 'stupid labour' that had been universal on our farms. Harvest was now just a few days' visit from a contractor with a combine. Grain was lifted by auger to the loft instead of being carried up the stairs in bags weighing more than two hundredweights, or sixteen stones, and the chaff was blown through to the chaff house or the cattle court rather than being carried in great hessian-wrapped bundles. John Allan was satisfied that, with the rise in farm servants' wages, 'the most expensive machine nowadays is a man's back.' But there were to be no more ambitions. Low would keep the land in the best of heart and keep his reputation for having everything ready for action on the spring day, and being last to start and first to finish at harvest time. James Kelman would continue to produce the cleanest milk in the county and year by year a little more milk from forty cows in the same byre. Low would have liked to put a roof over the silage pits for a Dutch barn and he would have knocked down the fine old byre that greeted William Yull when he arrived in 1837 to make way for a cattle court. He would have roofed the midden for more cattle courts, and James Low knew that grants were available for such works and that the government would pay for a lot of it. But the farmer wouldn't move. Low was able to do little things, however. He took to ploughing up a field of old grass before harvest and getting a flying start with the next spring's work by sowing a field of winter wheat. That was an astonishing advance. When John Allan came home from the war, seven quarters of oats was a good crop, but Low's winter wheat produced in its second year, fifty-two hundredweights to the acre, over fourteen quarters.

John Allan was inhibited by two things; the first was an unholy terror of debt. No doubt that had a lot to do with the way his grandmother and he had been left with £900 when his grandfather died, and that being all gone when he graduated to his first job on the *Glasgow Herald*. And the other was his health, which failed badly in 1956, and that had something to do with my family's long affection for strong drink.

John Yull, my great-grandfather, who enjoyed his drink so much that he needed three gallons of whisky to get through the New Year celebration in 1891, may have had the best of my family's long and intimate association with *usquebae*. There is no record of him being the worse of drink. If he did need occasionally to be lifted into his gig, it was because he was tired after a hard day's auctioneering. And if he had had enough drink to choke a horse, he would be sober by the time his great mare, Clatterin Jean, had taken him home through the night air.

Nor is there any reason to believe that my great-grandfather on the other side, Alexander Allan, was ever the worse of drink, though he bought his whisky in grey beards. These were china pigs holding five gallons, usually about proof strength − equivalent to about fifty-eight bottles of Macallan at 70 per cent.

Sadly, John Allan was not so lucky. All my life with them my parents had a drink with any guests that came in about, and when my Uncle Mike, Maitland Mackie the second, who finished as Lord Lieutenant of the county of Aberdeenshire, came for afternoon tea little would be left of a bottle of Glen Grant 10 year-old. I can still see my father when the visit was almost over, with the bottle in his hand and offering it to his brother-in-law with, 'We'd better just finish it.' And Mike answering with 'Oh well . . .' John Allan believed that the whisky bottle was the wrong size, 'It's too much for one, but it's not nearly enough for two.' But my parents also drank routinely when no one came. They would have a glass of Dry Fly Sherry before lunch and they would have one before their tea and they would certainly have a couple in the evening.

Despite that, I only once noticed that my father was not quite in control. He never staggered and he would sing at any time if he were happy, so if he had been 'singin fou' it wouldn't have been noticed. And he was a quiet man so 'roarin fou' just wouldn't have been him. But there was the time when we were driving home from the only riotous party of which I have ever heard being held at my grandparents' house at North Ythsie. The Old Folks, who, by the standards of their children and the people those children married, were virtually teetotal until Granny started to forget things in her eighties, held court downstairs. Their two noisiest sons, Mike and George meanwhile, were allowed an unlimited budget to host a fever of inebriation upstairs in the billiard room.

I was shocked to hear a conversation between my Granny and my Uncle Mike, 'Oh Mike,' Granny said, both hands flat on his chest, 'I've never seen you the worse of drink before.' That only showed how little of the time she had been around, but Mike was fully in command.

'I'm not the worse of drink, Mam, I'm the better of it.' Happy laughter.

We didn't get away until Mike had offered his parents a wonderful vote of thanks in which he reported that there was, 'A considerable body of opinion which thinks you should have a party like this every year – at least.'

When my parents and I eventually took the road in the old

Austin Ten, it was decided, for reasons I understand well now but which were a mystery to me at the time, that we should take the less public route home through the grounds of Haddo House. What was the use of that, I wondered. It being nighttime, we weren't going to see the pheasants and the roe-deer anyway. I remember being puzzled that my father should be driving so slowly, and wondering why the thick rhododendrons which lined the road through Lord Aberdeen's policies came swinging out to meet us before we burst gallantly through them to regain the narrow road.

'Why are you going so slow, Daddy? Why are you going so slow?'

There was no other conversation but my mother saying, 'Shut up, Charles. We'll soon be home. Shut up, Charles,' in a tense voice which wasn't my mother's usual.

But I don't remember her ever being the worse of drink. I remember her having a very red face and being in uproarious good form at a party at John Mackie's house in Essex, but apart from those two times I never saw either of my parents affected by drink. Perhaps it was because they were always a little drunk, for it was always there.

It had an effect. John Allan had had jaundice, which is a disease of the liver, when he was born, and we like to think that is why the drinking, which for success depends on a good liver, took such a toll. I was not much involved for I was away at my strange boarding school in Devon, but in 1956 John Allan was taken very ill. The ambulance came to Little Ardo and he was taken out on a stretcher. James Low saw him go, for they shared the farmhouse. When the ambulance had left he returned to the kitchen and said to his wife, 'That's the last we'll see o' John Allan.' Jimmy wasn't often wrong about anything, and he had seen many cattle and horses die, but he was wrong about John Allan. In fact my father lasted another thirty years. He was very ill. Cirrhosis of the liver was diagnosed, but he made it, and strangely, all he would say about his ordeal was that when they were carrying him out of the house he saw a tear in his grieve's eye.

I believe that, for they had become very fond of one another, a bond that grew stronger and stronger over the thirty years that they shared the farmhouse and the responsibility for Little Ardo.

One sad effect of his illness was that it was the end of John Allan's diaries. From the day he returned from the war he had kept a daily record of what happened on the old farm and on what the farmer had been doing. On 5 June 1956 he soft-boiled half a dozen eggs in the fermenting silage pit. On 6 December, 'The bull has knocked hell out of another trough. He should save his energy for his work.' And on 25 December John Allan, 'Roasted and ate the turkey. A bottle of red wine, the Sunday scandal sheets, and early to bed in a rather Christian spirit suitable to the day, having paid some long-waiting creditors and left myself very bare. But the cows are rallying round – 145 gallons. Maybe they deserved the turkey and the cheap red wine more than I did.'

But search as I have through the heaps of his papers, I cannot find a diary for 1957. It was a waste of time. I knew he would not have been able.

For John Allan's recovery was not complete by any means. He had only a small part of his liver functioning properly and the battle was then on – first to get the old man, who was only forty-nine, to give up the drink and then, more realistically, to drink less. My mother took ownership of the key to the press but never twigged that he had a key also. Then she didn't seem to notice the long study tours he made of the countryside. He was a regular at four general merchants' where, in the back shops, there were facilities for people paying their twice-yearly accounts to get a dram as a luck-penny. And if the accounts were already paid a farmer could get a drink in advance, or he could ask to buy nails, which was code for having a much desired dram as a business expense.

And of course the press isn't the only place in a house from which a man could drink. The first whisky I ever tasted was from a bottle of Abbot's Choice that I found among the cleaning agents in the cupboard under the stairs. It was so disgusting that I was sure it really was some chemical like paint-stripper and that I might have done myself a serious injury. I have enjoyed draughts long and short of all manner of drams, and even enjoyed a Japanese whisky that boasted that it was 'made from genuine Scottish grapes', but it is a legacy of that bottle in the cupboard under the stairs that I have never managed to hold down, let alone enjoy, Abbot's Choice. When my parents were leaving the old farmhouse for their retirement home, far from the disasters their son was perpetrating at Little Ardo, we found seventeen empty whisky bottles down the back of one of the free-standing bookcases in which he kept his thousands of books. And in 2007, when his granddaughter Sarah was doing up the little lean-to that had at various times been his study and his bedroom, and emptying the loft space, she cleared out eleven assorted spirits bottles, which showed Old John's catholic taste in strong drink.

That John Allan survived for thirty years after his serious cirrhosis attack in 1956 was undoubtedly due to the attentions of Dr Annie Anderson from Oldmeldrum. She was different. She was a strongly built woman with an Eton crop, who wore tweed suits and lived with a lady companion. She had a great sense of humour and admired John Allan. I remember her saying to my father, 'John, you are the only one of my patients who is the least bit distinguished, so I'll need to keep you alive.' And she set about doing just that. Her recipe was a huge boost of vitamins administered with what my father thought was far too thick a needle, and a daily diet of 'good-for-you-pills' that hardly left room in his stomach for food, let alone drink. And Dr Andy was a wise old bird. My mother had already tried to get John Allan to stop drinking and had failed, as had all his previous doctors. Dr Andy knew better than to try. She did a deal with the old man. If he submitted to her medical regime and ate all his pills, she would allow him two bottles of Guinness at or about bedtime.

The patient accepted the deal and my mother waited anxiously to see how it would work out. After a fortnight the good doctor was back. Yes, we all believed the patient was a bit better. Certainly he was a better colour. 'So how are you getting on with the diet, John'?

'Oh, nae bad, doctor. I'm more or less up to date with the pills', he said, and eagerly, 'but I'm miles ahead with the Guinness.'

He would have been sixty then and he stuck to the booze for another ten years. The doctor's compromise, as well as reducing his intake considerably, did provide the vitamins which may have helped to substitute for a liver, but even when my attempts to drag the old farm up to date had driven him away to Lincoln, away from his car and his string of general merchants, he still drank. In Lincoln they had a very swish mews cottage, and there he said he had all he needed. There was a delicatessen just down the street, the cathedral was just across the street (though I doubt the archbishop noticed a big increase in his business) and he was within very easy walking distance of more pubs than you could get round in a day – even starting early.

Now it came to pass that Jay Allan, the old man's grandson, went all the way to Lincoln, disturbing the busy schedule of a teenager at play, to see his grandparents. He was most welcome. He gave them all the news of the family and of the farm and of his hopes to be rich one day. By early evening the two old people, used to their own company with a fire and an endless supply of books and delica-tessen, were about exhausted. John suggested to the young man that he might like to go out for a pint. How thoughtful. There was nothing Jay would rather do than seek out the action in a new town.

'Now you should go left out of the house and don't go into the first pub. It is very rough. Go down and across the street to the next one. There's a very nice older couple in there, they keep a very clean shop and they'll look after you.'

It should have been obvious what Jay would do – head straight into the rough joint. There was hardly anyone in, for, while eight o'clock might be getting late for his grandparents, the night hadn't started for the youth of Lincoln. However, the barman was friendly and after giving him his pint asked Jay if he was working or holidaying. He had guessed he wasn't a new communicant at the cathedral. Jay said he was down visiting his grandparents. The barman's face lit up.

'Allan?' he said. He had recognised the accent and put two and two together. 'Proper gent, old Mr Allan,' he said. 'He's in here every night, five o'clock as regular as clockwork. Two double gins, swift as you like, and then away home for his tea.' Some things never change.

But the old man's love affair with the drink did eventually change. He had done his best work as a writer while sustaining himself with forty fags a day, booze and a gallon of the blackest coffee you ever saw. His excellent dairy cattleman, James Kelman, walking up to milk his cows at four in the morning, used often to see him with the light on in his little office, writing away. Kelman wondered if John Allan was up early or bedded late. The truth was that with his special diet of food, drink and nicotine, he often wrote all night. It was common for John Allan to write till he was finished or until he was done. He would then sleep until he woke. Have a fag and cough helplessly for about ten minutes, get the coffee on, take a small 'morning' and write on.

And then when he was seventy, without warning or any special encouragement, he gave them all up at once, completely and forever. I tried to praise him for the strength of will which managed so complete a conversion to a healthy lifestyle, even if he had left it a bit late. But he would accept none of that. 'I just don't want them any more,' he said. His passion turned to single portions of delicious creamy desserts from the deli. In a letter to an old friend he described his existence at that stage as, 'Staggering from pudding to pudding.' It seemed to work, for John R. Allan lived another ten years until cancer of the blood got him in 1986.

CHAPTER TWO

Adolescence in Aberdeenshire
1955–57

In 1956 I left the strange boarding school in Devon where the children called the teachers by their first names, bathed together in the altogether and never got beaten no matter what devilment they got up to. I arrived home to await the results of my General Certificate of Education, which I hoped would qualify me to follow my parents in a degree course to become a Master of Arts, probably in English. The sub-plan was that I would become a great football star and pay my way through university by playing for Aberdeen and become a Don.

Through all of our lives to that point, we loons on Little Ardo farm were dedicated to football. We all played at first, but while we all went to watch our heroes at Pittodrie, only the best players among us kept it up after school. That meant the two youngest Lows, Joe and Albert, and Ian Glennie, were still playing and we were all still keen. And when I came home from school my footballing career got a spectacular boost. Through a tip-off by Arthur Black the radio actor, I was invited to train at Pittodrie for a month. I know you will be expecting that I had to help the groundsman and got to have a kick-around in the in-goal area after the stars had gone home, but it wasn't at all like that. To my astonishment, I was treated like everyone else, while the manager, the great Davie Shaw who played full-back for Hibernian and Scotland, decided whether he had unearthed the next Willie Waddell.

I had to share the away dressing room with two of my boyhood heroes, Ralph MacKenzie, who was now playing centre-half for Raith Rovers, then a First Division team, and Don Emery who was now with Fraserburgh in the Highland League, where the man who nearly played for Wales as a full-back was now in a new lease of life as a centre-forward. Both of those had been playing when I went to my first match in 1947. Once changed we went out onto the track which surrounded the pitch and round which the policemen paraded during the matches. Round and round we went, strolling along the ends of the pitch and sprinting along each side. This went on for perhaps half an hour until everyone had turned up. Then we played a ninety-minute game, sometimes at Linksfield School, half a mile's trotting away, but usually on the hallowed turf of Pittodrie itself. This was a big step up from playing for Methlick Juniors or even Dartside Rovers in the South Devon Senior League. Bobby Wishart and Archie Glen were current Scottish internationals. Gentleman George Hamilton and Archie Baird, wartime internationals, were just fading, but both played at least once. And Harry Yorston, who had just shocked us all by announcing his intention to give up at the height of his career, and, despite being the darling of every susceptible woman in Aberdeen between the ages of ten and seventy, become a lumper (as the porters were called) in the Aberdeen fish market. And Paddy Buckley, who was well on his way to a record tally of goals for the Dons when he was retired by a cartilage operation which he didn't take seriously enough. And perhaps Aberdeen's second-best goalie (after Jim Leighton) was there, Freddy Martin, fresh from playing at Wembley when England beat Scotland by seven lucky goals to two.

It was something of a dream come true for the boy who had kept a scrapbook with the photographs of all those great men in it. And the dream didn't disappear as soon as the first whistle blew of a real game. I had a plan. I wouldn't try to be clever. I knew I couldn't beat any of these professional footballers, though I might manage occasionally to race them to the ball. So when I got it I would aim it at one of my side who was far up the field and boot it as near to him as I could. This was never a disaster, as at least we made ground and the very first time it was a complete success. I got the ball in the right-half position and banged it up in the general direction of Paddy Buckley. Now, as centre-forward Paddy was used to getting the ball with the centre-half breathing down his neck and trying to turn him or flick the ball round him. He caught my long ball on his chest and as it hit the ground flicked it round his man and, as the 'Green Final' might have reported, 'walloped the sphere into the onion bag past the custodian, who was rooted to the spot'. A little later I was standing at inside left. The ball came bouncing straight at me, and not quite in touch with the ball but haring after it was Harry Yorston himself. With the poise of an old pro I just helped it on past the defence and Harry raced through and scored in the left-hand corner.

I don't think Mr Shaw was watching either of those flashes of skill, for I am afraid they were only flashes. By the end of my month at Pittodrie I had lost touch with these practice games altogether. I trotted up and down, and even sprinted up and down, but the ball passed me almost entirely by. At the end of my month Mr Shaw said he hoped I had enjoyed it. He softened his rejection a bit by recommending I join Rosemount and play Junior football. That was after all the usual stepping-stone to the professional game.

But I wasn't ready for that. When I was fourteen I had played for the Methlick loons team that biked the fifteen miles up to Turriff to play the Youth Club there and had only lost 2–0. I had been approached by a scout for Deveronvale who wanted me to play a trial in the Highland League. I had promised I would phone him when I had finished school and was available, so I did and was duly invited to play a trial at Banff against Buckie Thistle that very evening. Now comes a painful bit of this memoir, and I don't know what to think of it even fifty-five years later. But Banff is thirty miles from Methlick and there isn't and never was a bus to it. I had been told about the game at two o'clock, and the kick-off was at seven, so there was no chance even had I had the money to make it into Aberdeen and out again for a sixty-mile round trip. And my father refused to give me a lift. He didn't say he was too busy, because he wasn't. He didn't say it was because he thought I should take Davie Shaw's advice and play Junior football first. He didn't say, 'If you want to play football you can get your own way there.' He just said no. Until very recently it never occurred to me to think that he was an unusual father that wouldn't give his son a lift to a trial for a step-up in his favourite sport. Even when I set off down the Big Brae to start thumbing lifts on the Fyvie road, John Allan didn't relent and I didn't think for a minute that he should. Now, I'm not so sure . . .

Lifts were hard to come by and I arrived at Princess Royal Park at half-time. In those days there were no substitutions allowed. However, our left back was injured and as it was only a friendly the Buckie management allowed me on for the second half. I was offered terms but didn't sign. The logistics of getting to Banff were so much worse than taking the well-worn route to Aberdeen to play Junior; besides, Arthur Black had told me that the Highland League was 'a way out, not a way up'.

But first I had one game for Formartine, who at that time were an amateur team and played on the Pleasure Park at Pitmedden. That was a bit like my first practice at Pittodrie. Everything went right in a way that it never did again. We were playing against Cove Rangers, who the previous year had got to the final of the Scottish Amateur Cup, so their players had played at Hampden. Anyway, the loon from Methlick whom the *Press and Journal* had reported having played a trial for Deveronvale last week, scored three lucky goals in Formartine's best-ever win against Cove, 5–0. I can only remember one of those goals. Their goalie kicked the ball out of hand and was ball-watching as it bounced towards me on the edge of the centre circle, which was about as far as goalies kicked the heavy leather balls we used in those days. I hit it first time and it sailed over his head into the net. Eddie Edmond, the Pitmedden millwright, whose obsession Formartine was, was dancing up and down the touchline. But Eddie's dance was short-lived, for the next Saturday the rising star was playing for Rosemount against Sunnybank finalists in the Scottish Junior Cup the previous year and participating in another record win – 4–0 this time. I remember both my goals. The first was a short goal kick which with my great speed I was able to intercept and wallop into the net. The second was after a corner. I rose high where the ball hit me on the shoulder and sailed into the net. So I stuck to Junior football, convinced that stardom was round the corner, but not far round it.

Time for a bit of sackcloth and ashes. Even after I had signed for Rosemount, which was a couple of classes above Formartine at that time, I turned out for the Pitmedden crackshots when available, and this brought me a rebuke from a very unexpected quarter. It stung me. My father might not take me to Banff for a trial, but Old Maitland Mackie came to watch me at Pitmedden and our grieve, James Low, also came to watch me play, and I was proud. There was another who was keeping a particular eye on Formartine's centre-forward. The opposing centre-half was a more accomplished player than me, but he couldn't match my speed and I was giving him a difficult afternoon. He decided that a bit of derision and intimidation might help.

Now, I had come across intimidation before, playing for Methlick against Strichen in the Buchan League. My marker at Strichen had spent the whole of the second half telling me about the hidings he would give me as soon as the game was over and advising me to get off the park before he broke my leg. I was really terrified. Luckily Albert Low, my pal from Little Ardo, who was not subject to intimidation and would just have felled the centre-half had he tried that with him, scored a stunning headed goal and we won the game. I was absolutely astonished when, after the final whistle blew my tormentor came over with a big smile and a hand outstretched and thanked me for a good game.

So I might have coped better with intimidation this time, but I had never faced derision before. My marker kept up a stream of unpleasantness. 'Ye think ye're funcy, div ye? Name in the papers. Rosemount. Oh my God! How did you get on with the Queen's Park scout? It's a wonder ye're nae wearin yer sark oot ower yer breeks like the rest o' them. Och you'll never get a game for them. You couldna kick an erse . . .' I did get annoyed by that and took most effective revenge. It was very cleverly done, though quite unpremeditated. After one of our attacks had failed and we were running back to midfield while the goalie booted the ball up the park, my tormentor came past. I trod hard on his toe and at the same time let out an outraged yell. The ref heard the yell and whipped round just in time to see the centre-half taking a swing at me. He was sent off and I was pleased.

But then I received the rebuke I deserved. I would not have been surprised had it come from my saintly grandfather, but I would have expected James Low, who in the 1920s had survived many a three-rounder in the boxing booths at the cattle shows and feein markets of Aberdeenshire, and who even took his jacket off at a Hogmanay dance at Methlick in the 1950s, would think this a funny incident which served the victim right. I could imagine us savouring my cleverness and the justice of it for years to come. But not at all. The Marquis of Queensberry rules in strange places. The grieve sought me out after the game and said, without a hint of a smile, 'Aye laddie, I saw ye. Ye shouldna hae deen that.' I expected him to add, 'but that'll larn the bugger', but he didn't. James was serious.

Sport then, was at least half of what I remember about my life between school and university; that, and studying for the damned biology exam I had failed at Dartington, and doing an English Higher at Geddes Irvine's crammer to fill my days. Of leisure, apart from the habit of sneaking into a cinema I learned from Aberdeen's professional footballers, who were nearly always to be found at the pictures in the afternoons, the highlight was the dancing.

All our first experiences of the dancing round the marquees all summer and the village halls all winter, was of course the Hogmanay dance in the Beaton Hall in Methlick.

But let me start where, for several years of my youth and many years for others, Hogmanay started and finished, in the kitchen at Little Ardo farmhouse. That was not in the front of the house, where Jean and John Allan would rather have put out the light than risk a visit on Hogmanay, but in the old farm kitchen where James and Isabella Low held great times, not just on Hogmanay, but especially on Hogmanay. When a great time of hooching and drinking and dancing got up during the year Mrs Low called that a 'Hogmanay nicht' to explain what fun it was. And it was.

In the early days, after a couple of sherries – which was all we were allowed – Albert and I would set off round the village 'seekin wir Hogmanay'. We chapped at each door and chanted:

> Rise up guid wife and shak yer fedders.
> Dinna think that we are beggars
> We're only bairnies come tae play,
> Rise up and gies wir Hogmanay.

With luck we would just get a sixpence and an apple to leave them in peace. But if they invited us in we only had one encore. We sang the same words, but this time to the tune 'The First Noël'. Then we might get a glass of port wine and a slice of cake.

One thing which is now little remembered is the tremendous number of whiskies we drank on Hogmanay in the 1950s. And what is even less remembered is that these whiskies were drunk out of nip glasses, which were little bigger than a thimble. The practice was that whenever you met someone you took a swig out of each other's bottle. And at each house you let everyone fill your nip glass by turns, toasted their health and then knocked it back in a oner. Well, you could hardly take more than one sip at such small glasses. And that's another thing that has changed. Nowadays nearly everyone drinks whisky diluted. But in the days of the nip glass there was no room for water. Now, it is not well known, but it is a fact that whisky diluted with water is more easily absorbed. As we drank those tiny nips neat, and as it was often quite a way from house to house, you could drink the undiluted spirit all night without falling down.

We always started in the Lows' kitchen with a dram out of everyone's bottle bar one. We thought it a bit over-cautious, but while Jimmy took a thimbleful out of our bottles, he kept the lid on his own. When Albert grew older and more gallous he took his father on about this, and called him 'a miserable auld bugger', though with a smile. He was told, 'Na na. You lads'll hae mair than enough the nicht. I'll see ye again when your mair sair needin it.' He would save our heads and his drams for another day. And Mrs Low also had our welfare in mind with her baking, especially and wonderfully her sappy dumpling.

That early session would be fairly restrained and it might be later before Mrs Low came out with her excellent toast. She was a lady who was as prim as her domestic situation on the farm allowed, and by no means one to swear or be profane except under the most provoking circumstances. It was therefore with a degree of coyness not common in Aberdeenshire kitchens that, at a certain stage, when things were heating up but riot had not yet broken out, she would stand up, hold her nip glass high and recite:

Here's to the hills and valleys aroon
Petticoats up and troosers doon.
Naething on earth is half sae sweet
As twa bare bellies gaun tae meet

– and down her whisky in one gulp and laugh at her daring. Among all the other blessings she brought to our old grieve I was able to guess from that toast that Isabella Rennie did not bring a cold wife to Jimmy Low's bed.

That is a flavour, though doing justice to a Hogmanay nicht in the Little Ardo kitchen is beyond my descriptive powers. There was no television, so it was only later that we got used to listening to Andy Stewart and letting the White Heather Club do the dancing. There was dancing though. And, in that, we were before our time. Twenty years before dancers started twisting and shaking without touching their partners, we danced to Jimmy Shand on the gramophone. And it was not organised square dancing but free solo dancing of a type not much different from what I saw in Ronnie Scott's Jazz Club in London about 1957. Albert Low and I were particularly active and one of our least orthodox steps was a short gyrating warm-up of maybe a couple of bars and then an attempt to kick the central light which was the main illuminator of the party. There might be an occasional birl or a waltz with a partner, but

16

mostly our dancing was quite like a crowd of modern youngsters with even less inhibition.

The Hogmanay Nicht at Little Ardo would either precede or follow or even both, the Hogmanay dance at the Beaton Hall. That would just be the same as the dances we went to at least once a week throughout the year in the village halls or in marquees before or after the Ellon Show, the New Deer Show, Meldrum Sports and the Methlick Flower Show. But if I tell you the format of those dances you will guess that the Hogmanay dances were just the same, but more so, and most folks left at or before midnight to be home in Granny Low's kitchen, not for the bells and Auld Lang Syne, but for the clock on the mantelpiece above the fire and our song:

A gweed New Year tae ane and aa,
And mony may ye see
And during aa the years tae come
Happy may ye be.
And may ye nae hae cause tae grieve
To sigh or shed a tear,
Tae een and aw, baith great and smaa,
A happy, gweed New Year.

The generation in rural Aberdeenshire that is in its seventies now used to rely on the dancing for recreation. Most of us went at least once a week to dances in the village halls and in marquees, the night before or after the Highland Games, flower shows and agricultural show. My friend in the village, Alice Mackie, says with relish that she lived for her weekends and the dances. She was one among many. Sadly, I only lasted perhaps three years before I was off to university, but they were iconic years.

As soon as we were big enough to pass for being old enough for the pub we gave up the quarter bottles of Four Crown port with which we had fortified ourselves for the ordeal to come. Then it was pub until they shut at 9.30. Girls were not welcome in most pubs in the mid-'50s in rural Aberdeenshire, and that was a good thing. If you had a steady girlfriend it saved money to meet her inside the dances. We drank McEwen's export or Tennant's stout with an occasional nip of whisky or a rum and pep. There were no pints in Methlick until much later. There would be Jock Paterson and Charlie Stanger, who, despite being less jolly in his later years, was several times elected the British Postie of the Year. Albert Low would not join us for more than a few minutes. He was a crack shot at darts and he would be busy playing, because they had a very strange system whereby the winner was bought a dram by the loser and the winner stayed on to accommodate the next challenger. There was a shelf high on the wall, in Methlick's Ythanview, and I've seen Albert, already a bit unsteady on his feet, but his throwing arm as steady as a rock, with drinks lined up on the shelf for his attention. Strangely, even if he had half a dozen rum and peps still to drink, he never seemed to need a help to drink them up at closing time; I think he thought that he should drink them personally and toast his vanquished opponents. And then it was off to the dance, where the girls had been filling-in time dancing with one another or with the sleekit devils who had gone in at eight o'clock to try to collar all the best talent. In the Beaton Hall at Methlick it was girls at one side of the hall and the boys at the other, with a great pack of unattached males forming up at the door, smoking and chatting in the slow, country way. It was the same at Pitmedden, where the great Eddie Edmond ran Friday-night dances in aid of the Formartine Football Club, to which he ran a complimentary bus all over the countryside, picking dancers up for eight and taking them home after the dance stopped at one o'clock in the morning, or half past eleven on Saturdays.

We always stood for a few dances in the male pack, chatting or just eyeing up the talent. You tried to look out one who appealed to you, but in my case it was more important to see one who might have noticed me. They would very seldom look straight at you and smile. That would have been great. But you would see them looking away rather than catching your eye. Then it was a question of waiting for a suitable dance. When you were young and just starting, that meant waiting for one of the dances you could do. Once you had mastered the entire portfolio it was a question of moving when the dance you fancied was announced, 'Ladies and gentlemen, take your partners for the modern waltz.' Then it was 'Watch this, boys,' to your pals and, with a slick of the hair or a straighten of the tie if you had one, you would have to walk the walk, right across the floor to the quines. It was not a favourite with any of the boys, that walk. Everybody was looking, and even if they weren't you felt that their eyes were upon you, hoping you would fall flat on your face or worse. The quines would nearly always say yes if you asked them up. It just was not done in those kindlier days to refuse. The only exception would be someone who was so drunk he could hardly make it across the floor or if he did could not be sure that he had reached the right girl. It wasn't that these country lassies were so very agreeable, but it made no sense to pay half a crown to go to a dance and then refuse to dance with whoever asked you. A much more common embarrassment was that you'd be halfway across the floor when someone else would beat you to it. Then you had to make a very quick and therefore unconsidered choice, or return, tail between your legs, to the men's gate. It was a very public humiliation, to be avoided if at all possible.

One pressure the young man from Little Ardo's farmhouse felt very strongly at the dancing was the pressure to speak. I wasn't shy of speaking to girls, and if I got the chance I was not embarrassed to kiss and cuddle them. But what do you speak about when you're dancing? That embarrassment formed the basis of my choice of which dance to do. The Eightsome Reel was so noisy and had so much movement in it that there was no opportunity for conversation anyway. The same was true of Strip the Willow.

The problems were the dances where you took hold of your partner's right hand with your left and put your right arm round her and placed your hand on the small of her back. The hand at the back was important, and not just because you could feel her bra strap. If you were a good dancer you gave your partner a strong lead, using the hand on her back to show her whether to go this way or that and whether it was slow, slow or quick-quick that you were up to. The left hand was known as the pump handle. Students who had learned ballroom dancing at Madame Murray's in Aberdeen could always be recognised because they worked their hands up and down like their grandmothers cranking the handle of the pump which brought the water up from the well at the back door. Those who had been to Bert Ewen's classes, which moved between the village halls or had taught themselves by asking a quine up and then trying to watch her feet, were smoother – at least, they didn't work the pump. Anyway, these dances where you were in close proximity to your partner were much feared because you could hardly avoid speaking to her, and what would you say for a whole dance?

The music was live, of course, and there were ever so many dance bands in the countryside. Methlick had its own, the Marrs, small farmers from Brainjohn just outside the village. Their line-up was typical. Mrs Marr led on the piano. Doug Marr played drums or fiddle. Laurence was a very talented accordionist and father Marr played the trumpet. Others played from time to time, but those were the mainstays. Normally they would play from eight till one, with one break for a cup of tea.

The programme was modern dances like the quickstep, foxtrot and modern waltz, and each of those could be a novelty which meant an 'excuse me' – you could cut in if someone was up with the partner you fancied. There were occasional ladies' choices, where it was the girls who had to make the embarrassing walk across the floor. I also found that embarrassing. You pretended not to be caring whether you were asked up, but of course you were, and it made me very nervous when I saw a girl doing the walk towards me. But was it me she was after? I think some of them did it to put their real quarry off the scent, but I quite often saw a girl looking at me as she hurried across the floor only for her to ask a bloke I was standing near to. Then there might be a double novelty, where the girls were allowed to cut in as well.

And there were traditional dances, including the Eightsome Reel, the Gay Gordons, Strip the Willow and the Dashing White Sergeant, as well as the waltzes: the Old-fashioned, the St Bernard's and the Waltz Country Dance. And those jolly clumping dances such as the Polka, the Boston Two-Step, the Highland Schottische and the Canadian Barn Dance.

The Eightsome Reel was very important – there could be as many as three Eightsomes. And they were almost like a competitive sport. In the Methlick version there was no tiresome figure-of-eight for addled brains to master. The eight bars where others do a figure-of-eight were totally devoted to 'birlin'. That is to say, linking arms with your partner or the lady opposite and spinning round and round as fast as you could for the full eight bars. Some got greedy and went on longer. The birlin was really most athletic. For many of the young men it became a thing to show off at. You swung your partner faster and faster until her feet came off the ground. It was surprisingly easy to maintain your balance and spin on at speed with your partner's feet at waist height and skirts everywhere. Best of all was to spin her not only round and round but also up and down, so that her feet would touch the floor at one side of her circle and be maybe five feet up at the other side. And it was competitive. While some quite liked it and most took it in good part, the quines did try hard to stay on their feet. For that they held their partner as far away as possible and leaned towards him, while the lads leant back and 'ruggit like hell'. And of course there was the odd occasion when one of the lads was swung right off his own feet and swung round like a quine. The boys didn't like that, but their pals did.

You can see why the Eightsome was more popular with the lads than with the lassies, but there was another reason, unconnected with the fact that a girl who was being swung off her feet might become detached from her partner and sent sailing and then scudding across the floor, collecting skelfs in her fleshie bits. It was the damage caused to bare female arms by male tweed jackets that made the eightsome a fearsome business for the girls in the '50s. There was great pressure on the inside of the elbows of the birlers and many a girl had arms that were red raw after a couple of Eightsome Reels.

The rowdy dances were easy for the speechless. But the take-hold-and-glide dances were very hard. And we made them worse by letting go with the pump handle, and with one hand round each other's waist we would walk round for a bit, doing the odd quick-quick or sometimes a rhythmical little quick, quick, quick – slow. You really did need to have something to say then.

I found it very hard. I should have listened to the girls who took the lead in conversation and were never stuck. 'Fit's yer name? Far div ye come fae? Foo did ye get here? Did I see ye at the New Deer Show? Did ye bide the dance? Terrible day again. I see there's been anither accident at Fumblin Neuk. They'll hae tae sort that road.' It didn't really matter how you started off, I now realise. If you can just get a start and listen to the replies and react to whatever's in those replies you can twitter away with anyone. How I wish I had known how to do it when I was at those country dances. In fact, I wish I could do it now. I find that I am all right with my own age group, but with anybody much younger, which is a growing percentage of the population, I'm almost as tongue-tied now as I was at the dancing back in the 1950s.

Anyway, I would have liked to take my choicest partners up for the slow and potentially intimate dances where you could get in about, like the Slow Foxtrot, or especially the St Bernard's Waltz, where the more daring girls might take, when advancing towards you, a bigger step than was strictly necessary and so make extra contact; but with having nothing to say I could only take up girls I knew well, and who I knew didn't want conversation but were content to dance. I could get on all right with them, because I could say nothing as well as anybody and I was a good dancer. Otherwise it was the Highland Schottische, the Canadian Barn Dance, the Boston Two-step, a polka or some other athletic affair.

A good Saturday night at the dancing was as many Tennant's stouts as you could afford, which wasn't many, a good band, getting 'set' (the technical term for getting an understanding with one of the girls that you would see her home – or even round the back and then onto the bus). Particularly cherished, because all the boys who watched you walk the walk across the dance floor to ask her up would see you, was a step outside 'for some fresh air' away from the sweat and tobacco smoke. The ultimate was if you got a 'ride'.

But in that the tongue-tied farmer's son from Little Ardo was never lucky. I suspect that that was the majority experience, and even for those who did get lucky, it didn't happen often. An exception may have been Tibber McBain, my charismatic friend who in 1950 I had saved from drowning in the Middle Lake at Haddo House. I always thought the virgins of Aberdeenshire owed me something for saving Tibber that day. He became an apprentice plumber and had no inhibitions. His approach was simple. Early in his relationship with a woman, on the second revolution of the dance-floor perhaps, he just asked the question, 'Div ye ride?' And he kept asking until he got a positive answer. 'I get a few skelps in the lug, but I get a lot o' rides,' he told me seriously and by way of encouragement. I think Tibber was only being kind, for I saw him getting set all the time and I never saw him get a skelp in the lug.

There is a great deal of mythology about the *après-dance* shenanigans, but there isn't much of it to which my experience relates. Like the lad who persuaded the lassie to take a turnie on the anvil at the Gight Smiddy while he was escorting her home from the Methlick dance to Woodhead of Fyvie, five miles away. By the time he reached the crossroads half a mile farther up the road, the lad was getting interested again, but the lassie made it quite clear that she was not interested and would not be persuaded. 'Och well,' said the lad, 'in that case I'm nae cairryin this anvil ony farther.'

My experience is closer to the lad who, having walked this lass a long way home, could not persuade her to intimacy beyond kissing and cuddling and said eventually, 'Well, there's nae use the three o's stan'in hear ony langer. Cheerio. I'll maybe see ye at the marquee next Saturday.' Not that I would have known how to ask the question in the first place.

CHAPTER THREE

Getting Motorised

1957

As loons on the farm we had enjoyed tremendous freedom by virtue of our bicycles. We all had bikes and we could go as far as we liked on them. At least we got to Oldmeldrum every year, on the third Saturday in June, to the Sports, which in most other villages in Scotland would have been called a Highland Games. That was a great day's adventure. There were clowns who rode motorbikes that exploded, high-wire acts, Highland dancing and all those athletic events I was to do so well in later years. There was always a celebrity who came to open the Games and I am proud to have been there when Richard Dimbleby opened the Sports and when his son Jonathan did so twenty years later. On one of my first visits to Meldrum with the loons I found my Uncle Mike was in charge of handicapping the children's races. I was very proud of that until he recognised me and put me back five yards because I was so much bigger than my age-mates. That was all right for him, and may even have been fair in a sort of a way, but it had considerable consequences for me. It was ten shillings for first place in the boys' race, which would have doubled my available cash for the day. But placed farther back and with about forty boys to barge my way through, all I could manage was second and five bob.

But as the teenage years crept up on us bigger adventures were suggested by the internal combustion engine. One of Albert Low's steps towards the freedom a car would bring was when he was allowed to get his father's car out of the old stable which was now its garage and bring it round to the back door so that Jimmy, having done his day's work, could get away to Pittodrie or to the bowling, which became quite a thing in the summer evenings. Methlick's green not being laid out until 1980, Jimmy had to make it over to Pitmedden to play, and a very competitive game he played too. Jimmy Low liked to win. And he didn't like to be late, so the minute or so that Albert saved him by getting the car out from the garage was important.

On this occasion Jimmy was due to play in a very important semi-final and with ten miles to go and his tea not ready till half-past five, time was tight. Albert went off proudly to the stable for the Austin Seven. Sadly he missed with the first turn of the starter, tried a good plunge of the accelerator the second time and flooded the carburettor. There was no way she would start without a push.

Now, Little Ardo is ideally situated for those with low batteries or flooded carburettors, so, in mounting fury, and with Albert pushing, he set off down the gentle slope towards the Big Brae. He tried several times to get a bump start but endured only failures until, down past the Single Hoose and onto the steeper hill, he finally got ignition. Jimmy slammed the old car into reverse and roared backwards up the single track road to the farm. It was a very nifty bit of driving, but when he reached the steading he swung the car into the long grass that separated the roads to the piggery and the rest of the steading for a one-point turn. It was not one of his better moves. The grass was very long, it being July, and that area being totally unfenced it could not be grazed, and it was never cut. There was a horrendous crash of tin on steel as the old Austin shuddered to a halt. The dykeside plough had been parked there for the summer.

'What stupid bugger left the ploo there?' Jimmy fumed, know-ing full well that that was where the plough always spent the summer, and if Willie Adie was a stupid bugger it wasn't because Jimmy forgot where he kept the dykeside plough.

His car didn't look good. Everyone at the bowling would see why he was late. He would have to go down a bit later for his paper for a few days in the hope of missing the discussion group which he knew would be majoring on his misfortune for some time.

25

Anyway, the mechanism of the car was unhurt and Jimmy got to Pitmedden just in time for his game, though he was cutting it fine. He needn't have bothered. His concentration was gone and he was not needed for the final. It was some time before Albert was required again for garaging duties.

Well, that was the grieve's youngest's first famous exploit as a car driver, but what of the farmer's son? Well as soon as he was old enough in law, he went into the car market and bought a car of his own. This does not mean that I was a spoiled brat. A brat yes, and in many ways I may have been spoiled, but not in the matters of getting all the best in toys and clothes and exotic holidays. My first car cost £25 and all my older and wiser friends said I was robbed. I preferred to agree with the 'friend', who disappeared without trace after the sale, that he had done me a great favour in selling me a delightful ex-Post Office van, a Morris Minor 1935. It had a wire mesh grid between the seats in the front and the letters and parcels in the back. What the grille was for I don't really know. It would have been handy a few years later when I was the farmer at Little Ardo and collecting calves for my bull beef enterprise. A grille like that would have prevented the calves from sucking your ears and otherwise distracting the driver. It would also have done well for transporting hens – at any rate I called my first car, the Hen Coupé.

I set about improving my buy. I didn't really like the Post Office livery. When people saw the little red van they were always stopping me and asking me to take their letters. And I got fed up of people asking me if I had failed at the university and started with the Post Office. I saw people giving me a queer look as though they thought I had pinched my treasure. So I decided to paint it. It had been a long time since Henry Ford had said he would provide any colour of car you wanted as long as it was black. In 1957 there were brown cars, green cars and maroon cars that were almost brown. But those colours were far too tame. From my mother's treasure trove of half-used pots of paint under the stair, I was able to get enough sky-blue paint to do the bodywork. Then, reminiscent of the day, in 1943, when I had made such a splendid job of painting my tricycle and the front step and myself, and despite my clear memory of the thrashing I got for my endeavours, I got a pot of shocking pink for the mudguards. With The Hen Coupé emblazoned on each side in black, that was the aesthetic side of my first car attended to.

Then I had to do a bit on the engineering side. I noticed that, whereas most cars sat square on their chassis, mine seemed to be down on the driver's side. So I jacked her up for a look. Despite knowing not a thing about the engineering of cars I was able to sort the problem to my satisfaction. In 1935 the Morris Minor Post Office vans were suspended on two large springs consisting of about half a dozen leaves of steel about a quarter of an inch thick bound together. At each end of each spring was a strut about three inches long, and from those four struts the car was suspended. Or it should have been. In the Hen Coupé's case one of the struts was missing, so that the body of the car was free to come down and rest directly on the spring. Nothing daunted, the young mechanic got hold of a log from the stick pile at the back door, being careful to select one near as damn-it three inches thick. With several yards of electric fencing wire, I lashed the stick onto the spring and lowered the body of the Hen Coupé back down. It was now squared up and looking well. The mechanic was proud and saw no problems of safety in his repair . . . but think of the money he had saved by not giving the job to a garage.

Buying a car had been a ridiculous extravagance for a young man with ten shillings a week pocket money and no other means of personal indulgence except the money earned stooking and building cartloads in the harvest field. But this was expense with a purpose. In my last year at school in Devon I had had a wonderful summer playing cricket up to five times a week. And I hankered after a return to play for Dartington Estate and for Staverton, the tiny village beside the River Dart with its magical post-match debriefings at the Sea Trout Inn. We had had a very competitive school team and Ray Lance, the bursar, aided by Les Markham, a tough Yorkshireman who was head gardener on the estate, managed to get us together again for a total of three summer tours. For one week we would live in the school and play on the excellent cricket pitch there, and then we would set off on tour. Over three years we toured Yorkshire, Gloucestershire and Kent, playing on grounds that varied from the tiny Cotswold village where the slope was so great that when the fast bowler was starting his run-up to bowl uphill, the batsman could only see him from the waist up, to professional grounds at Harrogate and Canterbury.

All Games were friendly, of course, but not that friendly, as this will show. We were playing against a ground-staff team at Canterbury St Lawrence, the county ground made famous by Godfrey Evans and the fact that there was an ancient tree growing on the playing area. We just had to play around it. We batted first and the opposition intimidated us by the speed of their bowling, the accuracy of their throwing and by the fact that the wicket-keeper stood so close up behind our batsmen that his head was actually in front of the wicket. We were not doing well and losing wickets cheaply when our competitive spirit showed itself. Or at least Geoff Markham, a very cheery tough nut who had shared my room during our first year at the school, revealed his. Geoff waited until the bowler was about to bowl and lifted his bat to signify that he was not ready and turned to the wicket-keeper, whose head was a good foot in front of the wicket – quite against the rules – and said quite quietly, in an accent which retained some of his dad's Yorkshire vowels, 'If you don't stand back I shall knock your fookin 'ed off.'

Well, he didn't knock it off altogether and he didn't do it immediately, but the first time a ball was bowled wide on the leg side Geoff swung at it and missed, but his follow-through carried the bat right round and clunked the wicket-keeper on the head. I don't remember exactly where he hit him but I do remember the red bits on his face and the fact that he left the field. After that nothing could stop us. We had got the taste for blood. We recovered and scratched together something of a score with the driver of that terrible sky-blue and shocking-pink van contributing seven runs. Then we fielded like boys possessed and got them all out for a pittance. I don't really think you should hit people on the head with cricket bats, but it was, for me, a great sporting moment – if not very sporting.

So that was my mode of transport for the trip to Devon in 1957 for my first tour with the Dartington Hall Old Boys. And it was some safari. There is no way in the world the van would have passed an MOT had it been required in 1957, but what I didn't know was that the Hen Coupé would be subjected to a very thorough inspection by the Alloa police, though that was much later. It had a record speed of almost forty-eight miles per hour, but there were few roads on which the Hen Coupé could get to forty. From Little Ardo to South Devon was all of 700 miles for a start, and remember, in those days there were no motorways other than the Preston bypass, which provided a bit of clear motoring in the north of England. Otherwise I had to go into every town and village all the way to Devon. I had to wind my way down into Stonehaven and out again up the steep winding hills, which the Hen Coupé didn't like at all, though I just managed to stop her boiling. Then it was on through Fordoun to that great long street which was and is Laurencekirk, round the hairpin at North Water Bridge and into Angus. There the roads seemed very grand, but I was hampered by crawling through the great metropoles of Brechin, Forfar, Cupar Angus, Burrelton and Perth itself.

I guess you get the picture. It was the same all through Fife to Stirling and then through the vortex of Edinburgh. It took more than an hour to clear the capital and then it was more or less a smooth journey down through Haddington towards Berwick and England. By this time, twelve hours into the journey, even that callow youth was getting tired; besides, the light was fading and those on the Hen Coupé were not suitable for night driving. One day into the great trek and I still hadn't cleared Scotland. But where would I stay? Aha! My journey had been carefully planned. After a couple of pies, three pints and a lot of teasing from the locals in a roadside pub a bit north of Berwick, I took my leave and went to bed in the Hen Coupé. In his 1939 book about his travels in England (*England Without End*), my father wrote that he would rather have made his tour in a four-poster bed which would be large enough to hold a small barrel of brandy. He would be preceded by a brass band and would dispense shots of brandy to the eager villagers who would turn out to greet him. Well, the Hen Coupé was hardly that but she came quite close. There was a bit of mattress in the back where the letters would have been, and the sleeping bag which I had to take with me on the cricket tour, for the tourists were to live under canvas. The young athlete could stretch full out if he lay with his head at the back door and his legs through the hatch and draped over the passenger seat. There was no brandy but it was nearly ideal, and it was quiet in the car park.

After eating the pie carefully laid in for the job and swilling it down with a bottle of lemonade, I set off again at six the next morning.

You might think the next international landmark would have been Newcastle, but you'd be wrong. The destination was South Devon, so at some stage I had to get across the country to the west coast. The details and the reasoning are lost to me now, but suffice it to say that well into the second day things took a decidedly alarming turn for the colourful traveller. He was on that long straight road that leads south into Carlisle when, something like five miles out, all the traffic that is now on the motorway started to pile up in a great queue to get into and through the border town, with its city walls and narrow gates. That wouldn't have been too bad for me as the traffic was never stopped for long and I couldn't go very fast at the best of times, but things suddenly got a lot worse. The traffic in front of the Hen Coupé slowed down and came to a stop. I applied the brake and slowed but then, when I wanted to stop, I found that I had no clutch . . . not a bad clutch . . . not a slipping clutch . . . *no* clutch. I tried to stop with the brakes alone but the brakes being quite weak, and the engine being well advanced, in other words revving gently, I was unable to do so. I was chugging towards the car that had stopped in front. At the last minute desperation produced a stroke of genius – I switched off the ignition and shuddered to a halt just in time.

But, when the traffic started to move, how was I supposed to get going again? Well, I had a decent battery and I found that if I switched on the ignition with the Hen Coupé stuck in gear as she was she would shudder to a start and I'd be all right until I had to stop again. I proceeded like this for a couple of miles or so until I began to get even more desperate. What was I going to do once I was in the traffic jam which at that time was the old town of Carlisle? I would need to stop every few yards then. I would have to conserve my battery. The next time we set off I held back until there was a big gap to the next car. Then I drove very slowly to conserve as much as I could of my gap. Then when the traffic all stopped I undertook the stationary cars by putting one wheel up on the pavement and waving wildly as I passed them in a vain attempt to persuade them that I had no option. They did not understand and my cricket tour progressed to a chorus of horns, abuse and obscene hand signals from my fellow travellers.

Eventually, now well into the suburbs of Carlisle, I saw in the distance a sign proclaiming Shell-Mex petrol. There were few dedicated filling stations in those days so this must be a garage. My hopes rose and rose until with one last dash of undertaking I fully mounted the pavement, crossed a little strip of grass and bumped down onto the garage forecourt. And that was more or less it. There was a mechanic there. He took a brief look and said that the clutch cable had come loose. He stuck it on again, charged me ten bob (cash) and I was on my way again.

The rest of the trip was uneventful and really all I can remember about it was being charmed by the Cotswold villages of Stow on the Wold and Bourton on the Water. I arrived safely on the third day.

A great tour was had, with twelve cricket matches; Julian Bream, the classical guitarist, played in the evenings at the Dartington Hall Music Summer School, to which the cricketers were honorary guests, and for the first time (of three so far), I made myself sick with drink. Of course that over-indulgence wasn't my fault. Heavens no. My best pal from the schooldays, Nicholas Johnson (of whom more in Chapter 4), was on the tour as scorer and court jester. Somehow he had brought enough money to stand us a splendid dinner at the thirteenth-century Cott Inn. I remember we finished the meal with some showing off at how we could knock back the nips neat. But the real damage was done by Geoff

Markham, again you see, not my fault. The three of us and perhaps one other ended with a very jolly session in the White Hart Club at Dartington Hall's fifteenth-century manor. Sadly, the staff had also been partying and forgot their duties. So when it was his turn Geoff asked what we wanted and then liberated a whole bottle of it from the bar. I don't know how many such bottles we had but it wasn't one. No wonder that after a successful dance I was sick as a dog under the handkerchief tree in the historic Hall gardens. Like generations before me, I failed to learn the lesson offered by this experience, though I now knew that, wonderful as drink tasted going down, it was a disaster on the return journey.

All too soon the three weeks were up and the Hen Coupé and I set out for home. Starting from Kent, I just sailed straight up the Great North Road. It must have been on the evening of the second day. I was getting fed up of living in the Hen Coupé, which was beginning to smell as though it had been home for a long time to a young athlete with no access to laundry or bathing facilities, and I wanted to get to Little Ardo. I battered on through Scotch Corner to Newcastle and up by North Berwick to Edinburgh. Then up towards Stirling before crossing the Kincardine Bridge. It was there that the athlete met his Waterloo. The light was just beginning to go and I remember putting on the lights of the Hen Coupé and peering at the road, for I was on unfamiliar ground. Then, just past the Police College at Tulliallan, I saw a big road sign which proclaimed Alloa to the left and Perth to the right. Knowing that Perth was on the way home I took what I thought was the right-hand fork and bumped round this gentle left-hand bend. Suddenly, on what I thought was my side of the road, a big black car appeared. He jammed on his brakes and stopped. I did my best with my brakes and the Hen Coupé shuddered a bit and then rammed him. I knew by instinct that this would be my fault and I was quite right. A very angry man got out of the back seat and shouted, 'Do ye know who I am? I'm the Superintendent of Alloa Police.'

Not only had I chosen the wrong person to run into, but I had mistaken a big roundabout with a high bank in the middle for a fork in the road. It was not my finest hour. The local police were sent for. I must have made a good job of being contrite for the superintendent began to calm down and even told me a bit about his biography. His wife, who was with him, as they were on their way to some lavish night out, was a bit shaken and none too pleased that so much of her night out was being taken up with a road accident. Then they softened a bit more and admitted that they were surprised that this roundabout had in fact been the scene of a large number of accidents, though they couldn't really understand why. Could I explain 'off the record' why I had come the wrong way round this roundabout when I had successfully negotiated every roundabout from Methlick to Plymouth and back through Kent? Well, I didn't know what he meant by 'off the record' but it certainly didn't mean I couldn't incriminate myself by what I said. Already in a hole, I started digging furiously. It was just a question of how may charges they would bring against me.

I explained that it was getting dark and my lights weren't too good. That was my first charge. I could see the road all right, but they didn't shine widely enough to illuminate the road sign and I didn't see it until the last minute, and there was a bit of a Scotch mist and my windscreen wipers didn't work unless I put my hand outside and waggled them manually. That was another offence. But really I thought it was that I had been driving for thirteen hours with only the briefest breaks for crisps and lemonade, and I was tired. Well, 'off the record' or not, that was three charges, including driving while incapable through overtiredness, to add to ignoring a traffic signal and going the wrong way round a roundabout.

When I was questioned by the boys in the squad car it was 'Did you sound your horn?' 'No, my horn doesna work.' (So I was charged with having a faulty horn: charge no. 6.) 'What speed were you going at?' 'I couldna say. My speedo's never worked.'(That was charge no. 7.) Then they went into the roadworthiness of the Hen Coupé. First, the brakes were tested on the road. They set a brake meter on the floor and it gave a reading on how much braking it could detect with first the hand- and then the footbrake hard on. But the cop who was driving just didn't have the hang of my van. He hardly registered any braking effect at all. I persuaded him to let me drive and then the Hen Coupé just passed. But the police wouldn't accept that. Apparently all road vehicles have to be foolproof – any policeman has to be able to drive them – it is no use knowing how to stop the car by pumping the footbrake and leaving the clutch out for the first second or so to get the braking effect of the engine. So faulty hand- and footbrakes made charges 8 and 9.

I have forgotten three of the misdemeanours with which I was charged, but I do remember with absolute certainly that there were thirteen charges in all. All but one really referred to the safety of the vehicle and yet, after dark, they allowed me to proceed on my way up to Aberdeen and back to the farm. It had not been one of my best days and yet I had acquitted myself much better than might have been expected. I had gone and apologised to the super-intendent's wife, 'For spoiling your night out', and she had clearly been charmed by this gesture and my obvious contrition. So much so that she invited me into her car for a chat and to tell her all about where I had been driving this extraordinary heap. Anyway, when she and the superintendent of Alloa Police finally took their leave of me, she said, with a very warm smile, without a wink but in a conspiratorial way, 'I'll do my best for you, son.' I think she must have done, for of the thirteen charges eventually laid against me, only two were brought – ignoring a traffic signal and faulty handbrake. My total fine was £6. I remember thinking that while £6 was a lot for those two minor offences, it wasn't a lot for hitting the superintendent of Alloa Police and the thirteen original charges.

Even with the departure of the superintendent and his lady, I wasn't finished. I think it was because they were intrigued by this heap of scrap they had picked up on the road, but at any rate police insisted on taking the Hen Coupé to their garage in Alloa to see if they could find anything else wrong with it, or perhaps just to understand how such a basket case had done 1,300 miles in the last three weeks.

I was given a cup of tea and a seat while they did whatever it was they did. After maybe half an hour I was called in to explain my van. She was jacked up, so, really for the first time, I had a proper look at her. They called me to 'look at this'. I thought I knew what 'this' would be. My expert repair of the broken spring would have caught their eyes. I have no doubt at all that it had, but they played cat and mouse with the poor student. The mechanic pointed at the other spring. The smallest of the leaves had a tiny chip out of it. 'That could be dangerous,' he said, while I tried to stand so that the other spring, with the log from the woodpile and the neatish job I'd done with the electric fencing wire, was behind me, in the ridiculous hope that they wouldn't see it. Anyway, they convinced me that they hadn't seen the major problem and I was very relieved when they just added 'broken bodywork' (charge no. 10) to the list of charges.

It was about midnight when they let me go and I set off on the last leg of the journey home. Being like the snail who as well as being slow, carries his house on his back, I had no need to drive any farther than I wanted to. On the other hand I didn't want the same policemen who had spent so much of their time going over the Hen Coupé to find me trespassing or illegally parked, so I drove up into the Stirlingshire hills for an hour or so. There I found a nice broad verge, where I parked up and slept the sleep of the dead until the birds woke me some five hours later. Then it was all speed for home, stopping only for lemonade and three bradies (a North-east bridie) at the shop on the corner at Kinross. There were all sorts of alarms and strange noises from the gunwales of my gaily coloured barge, but all went well. In the late afternoon I chugged up the doctor's brae at Methlick, up the Loans and round the corner at the Cottar Houses. When I put my foot down to ask for one more burst of enthusiasm from the gallant Morris Minor, she cut out. I nicked her out of gear and free-wheeled down to and though the steading, and left her where she stopped, in the grassy bit in the close beside John Yull's well, where James Low had crashed into the dykeside plough, and the broken farm machinery lay waiting for Mr McConnachie's the scrap merchant's next visit.

Had she landed in the possession of a mechanical person I have no doubt the Hen Coupé would be starring today at vintage car rallies, restored to the livery of Her Majesty's Post, and with a new bracket holding her spring firmly in place. But I had had enough. I had no money. Car ownership would have to wait. Mr McConnachie gave me a pound for the scrap and Mrs Low gave me twenty Capstan Full Strength cigarettes for the travelling rug which had come with the car and had been such a comfort to me on my 1,500-mile summer safaris.

It was a move from the ridiculous to the sublime to go from the Hen Coupé to my grandparents' Austin Princess, or Sheerline as the makers preferred to call her. Old Maitland and Mary Mackie were not conspicuous consumers. Their previous car, which had seen them through the war and well into peacetime had been a Riley 9 horsepower. Heavens, it wasn't even as big as the Little Ardo Austin Ten, so it was a huge step up to the six-cylinder 2.2-litre Princess. The aspiring student was quite excited when he was invited to go to North Ythsie and take Mrs Mackie shopping. The old lady who had looked after me during the war had never fully recovered from breaking the head off her leg joint when she was seventy. She got back behind the wheel, but at eighty-one she was reckoned to be past driving to the shops at Aberdeen and even the short trip to Tarves was becoming hazardous. It is only a mile and a half up to Tarves, but I was looking forward to learning to drive such a grand car. I also liked my granny and was flattered by the chance to do something for her, who had done so much for me, and even forgiven me for eating all her show gooseberries in 1949.

We arranged ourselves in the splendid gleaming status symbol, she calm and elegant, I proud and just a little anxious. Despite the fact that the controls were quite foreign to me, I managed to get the engine started.

'Now Granny, what about the gears?' I had hoped to be given a run through of the four gears, one more than I had had in the Hen Coupé, and then there would be the knotty question of reverse. Where was that?

'Oh just put it in there, dear,' said my driving instructor, jabbing the gear-stick towards two o'clock. I found a gate there and with a generous boost on the accelerator the car took off. Sadly, it was not a stately Princess that shuddered and bounded round the house and through the close with a nervous 'putchitty, putchitty, putch'. When we got up to thirty miles per hour or more our progress smoothed out and the young driver was now ready for more speed.

'Where do we go for top, Granny?' I asked.

'Oh no, dear. You just leave it in there and if you want to go faster just press the pedal harder.' The student was appalled. It was quite clear that the old lady only knew of one gear and that it was probably third. And yet she had driven the car for years, including to Aberdeen, where she did her shopping on Union Street.

'But what about reverse, Granny? What do I do when we get to Tarves and have to turn for home?'

'Oh no, Charles. You just go round the square and that turns you round.' So no reverse then. Well, by the time we got back to North Ythsie I had worked out the four forward gears, but I still had no idea where the gate for reverse was. Luckily I didn't need it. I don't think Granny did either, I don't believe she drove after that.

CHAPTER FOUR

A Gentleman's Education

1957–60

When my father went to Aberdeen University in 1924, the first of any branch of our family to make it to the Elysian Plains, he said it was the most natural transition in the world. He was not the traditional Scottish poor boy who set off from the croft with his bag of meal and half-barrel of herrings to experience the wonders of the big toon and life in a garret. He was poor all right. His parents and all but one of his uncles and aunts had all fled Aberdeenshire for the New World, so, when his grandfather died, there was no one left to take on the family farm of Bodachra and maintain the tradition of tenant farming.

His granny got a little fisherman's cottage by the Brig o' Don. While they lived in the but, they had to let out the ben to a student to make ends meet, while John was promoted to the roof space, where through the unlined slates, he could feel the snell wind off the North Sea and dream of a life on the open waves. So the boy had already enjoyed salt herring and was already living in the garret when he got into the university. The University of Aberdeen was just a short walk away over the Brig o' Balgownie, so he described matriculation as being 'as though I opened the back door of the cottage and walked straight into celestial fields'. He was already well-read, having filled the loneliness of his youth with books, and he was eager for knowledge.

For his son, going up to university was different. Brought up by intellectual parents who were both graduates, I was used to the idea

38

that education and books were important. I wanted to get a degree at the university and equip myself for a life of journalism at the *Glasgow Herald* – there was never any question of a farm for me until after my parents were finished with Little Ardo. I would do English, because journalists need to be able to write, I would do Philosophy, although I knew not of what it consisted, because that would help me to dazzle, and I would do Economics, because I had heard my mother dismissing Churchill as a peace-time prime minister because he had no knowledge of Economics. I didn't want to be dismissed, as a journalist or anything else, because of a lack of knowledge of Economics.

But my main concerns in going to university were far different from all that. I would get my degree, but my priorities were sporting and social.

It may have been a bad place to start, but I began with a complete survey of all the pubs in Aberdeen. My cousin, Maitland Mackie the Third, was the Chairman of the Students' Charities Campaign and he gave me the job of Beer Convener. The theory was that we needed a lot of bottles of whisky for raffles, and crates of beer to give the hard-pressed charity workers a refreshment after their day's work. My job was to go round the pubs begging for free drink. I visited every licensed establishment in Aberdeen – and there were over 130 of them. My routine was that I went in and ordered half a pint of light beer, which was of course dark in colour, the cheapest drink you could get, and the weakest apart from lemonade. I then asked if I could see the manager. Looking back over half a century, the reaction of these hard-working and often hard-drinking publicans to this lucky youngster who was living the life of Reilly at the university and en route to a life in the professional classes seems incredibly warm. In almost every pub I got either a crate of bottled beer or a bottle of spirits. What a gentle time it was – I never showed them any identification nor did they get, so far as I remember, a receipt, and they certainly did not get a letter of thanks. I am sure if one of my grandchildren tried it today he would get very little booze and however many skelps in the lug it took to make him give up.

39

Another of the things that defined my early days at the university was the disgraceful behaviour that led to my arrest and being charged with jail-breaking. When I was a lad, and even when I went to the University of Aberdeen in 1957, 'student' to someone in rural Aberdeenshire, at least, meant a young person in fancy dress behaving outrageously and rattling collecting tins in their Easter holidays. We were collecting for the support of the hospitals in Aberdeen, a leftover of the days before the foundation of the National Health Service. As well as tomfoolery and the collecting tins, there was a huge torchlight procession along Union Street, to which much of the population of the Toon turned out, and the Students' Show, an annual revue at His Majesty's Theatre, where the 'Scotland The What' trio and June Imray, 'the Torry Quine', made their débuts. To publicise those events, as well as for the sheer hell of it, there were also stunts, and it was one of those that got me arrested. It was nothing like as bad as the stunts undertaken by my pals Kenny Grassick and Bill Sey. Among their exploits were putting a chamberpot on the very top of the Mitchell Tower at Marischal College . . . and it's almost 270 feet high. They also dared the sheer drop of 500 feet by speeling hand over hand across Rubislaw, the biggest open quarryhole in the country, and left a banner in the middle as proof. They also replaced the royal standard at Balmoral with a skull and crossbones. That made the national press, but it was nothing to their next plan. They would go over to France, climb the Eiffel Tower and put an Aberdeen Students' Charities flag on the top where the Tricolor usually flew.

So over the two heroes went with their climbing gear and bedded down on the banks of the Seine for an early start in the morning. Not being great readers, especially in French, they had missed the story that, only the day before, some Algerian nationalists had tried to blow up Paris's great monument. That meant the French police were particularly jumpy about their tower. So when Bill and Kenny, both dark-haired and thoroughly swarthy on account of their passion for mountaineering, came sauntering along with their ropes they fell immediately under suspicion. They had got only about five feet off the ground when they were surrounded by a couple of hundred excitable-looking policemen with machine guns, and marched off. But even the French police seemed to have heard of the Aberdeen students and their stunts for, far from the guillotine, they were freed unharmed and even unfined, which was just as well, or they would never have had the money to get home.

Other stunts included the kidnapping of local dignitaries, who would be held for ransom. Top footballers would be kidnapped and ransomed under the threat that they would be held until after important matches. Stars of stage and screen would also be used as fundraising materials. With those events the stars didn't seem to put up much of a fight, which was undoubtedly because in most cases they were just as keen on the publicity as their captors. Even I, only eighteen, timid and law-abiding, was involved with Grassick and Sey in a kidnap, not for ransom but to gain publicity for our raffle. The victim was my schoolfriend, now London University student, Nicholas Johnson, who was later to become well known as Bertrand Russell's private secretary and a founder member of the civilly disobedient anti-nuclear group, The Committee of One Hundred. The story was that he hadn't sold enough raffle tickets and we kidnapped him in his pyjamas, sentenced him to imprisonment and set about delivering him after midnight to Craiginches Prison. Without so much as a risk assessment, I carried Nicholas on my shoulders up a ladder mounted on one of the Mackies' Aberdeen Dairy's lorries beside the prison wall. Fifty-two years on, in preparation for this volume, I went to Craiginches (now HM Prison Aberdeen) to make an accurate guess at the height of the wall. Even now I don't like to phone up and ask the warders. It is an awesome sight and not less than thirty-five feet high. June Imray, the Torry Quine herself, who became famous as the first announcer on the BBC to broadcast from London with a 'regional' accent, stood guard. Sey and Grassick, both sadly dead already, used their skill as mountaineers to ensure that Nicholas could be lowered gently to the other side and the rope then retrieved. The job done, Nicholas was left to his fate.

Our getaway driver, Maitland Mackie the Third, had some

difficulty in getting the old lorry going and we faced a Keystone Cops moment – as we were trying to get off a Bobby appeared and gave chase – on his bike. Just in time the lorry fired up and we got clean away.

But we proceeded to throw our triumph away, behaving as no hardened jail-breakers would have done. It was now far too late for June to be out on the town and she asked me for a run home. I was delighted, because she lived near to the prison and I was beginning to worry in case my friend would die of exposure in his pyjamas. What if no one in the prison heard his pleas for help? I borrowed Maitland's mother's Ford Popular car and managed to persuade June that we should go and have a listen for Nic's distressed cries while pretending to be a courting couple. The scheme wouldn't have imposed any hardship upon me, but sadly it fooled no one. We didn't even have time for a kiss. No sooner had we stopped beneath the wall than we were blocked in by police cars front and back. June and I were arrested and spent the night in Lodge Walk – a cell apiece. It was not a comfortable night's rest. I sat in a tiny cubicle on a narrow board which wasn't broad enough for any sort of rest. I often wonder if that is a bit like the 'stress position' they talk about our authorities using against terrorists; certainly, I got awful cramps in my legs. It was almost unbearably warm and stuffy. We were interrogated, charged with jail-breaking after they had explained that the crime of breaking in was the same as for breaking out, and the likely sentence the same also. Then about ten o'clock in the morning we were told to go home. The first thing we did was to read about our exploits in the *Press and Journal*, who were in on the plot, and to see the picture of me climbing that ladder with Nicholas on my back.

I don't know what to make of it fifty-two years later. The charges were dropped but it certainly wasn't all good. Poor Mr Imray got a terrible shock when he was wakened at seven o'clock in the morning by the police and told that his lovely daughter, who was doing so well at the university and would graduate soon, was under arrest. And we were lucky that no harm came to our victim. Nicholas Johnson had had the greatest difficulty in getting anyone in the prison to take any notice of him. He was cold and miserable and stumbled about shouting for help. The first responses were a few sleepy and then angry voices from fellow convicts telling him to give them peace to sleep, but he was eventually rescued and locked up. He was also let out in the morning without any charge. The prison governor wasn't pleased at the time, but Maitland Mackie met him socially a few years later when he said that in the end he was well pleased by the incident. We had shown that security wasn't sufficient at the jail, so he had got an extra warder.

I shudder to think what would happen to our grandchildren if they did now what we did in the 1950s. Health and Safety would have a field day. The lorry was not licensed or insured for use as a getaway vehicle. There was no risk assessment. There was no standby medical presence, although Grassick would soon be a qualified doctor. But those were gentler days, when the people of Aberdeen just smiled and said, 'Students! Fit'll they dae next?'

I won't bore you with it all or even the half of it, but when I said I intended to do a lot of sport at the university I meant it, and I did it. I still had hopes of a career in professional football, although the certainty that it would happen one day was fading. When Mr Shaw didn't come up with any sort of offer but advised me to play Junior for Rosemount in the Aberdeen Junior League, I took the advice and scored quite a lot of goals. I got a game for the Junior Select and most of the others in the team made it into the professional ranks; but when the papers kept saying Queen's Park were watching me I used to despair – Scotland's only amateur club wasn't going to help me pay for the beer. At any rate, I continued to play Junior football but joined in as many of the other sports as I could. I soon became captain of the basketball team, scoring a record thirty-four points in a game against a very poor Post Office Eagles team. I also played cricket, as a regular in the first team, though I never did very well. I never got the hang of the slow Scottish wickets, where you had to hit the ball hard to score. I had been used to English wickets, where the ball came at you so fast you had only to glance it and the ball would race away to the boundary.

Then there was the rugby. I was dazzled, not by the game but by the great men of the university who played it. Several, though still students, were in their last years of study and would soon be doctors like Kenny and Brian Grassick. And Johnny Moffett, who would soon be Junior Forest Officer Moffett with the Forestry Commission in Northern Ireland and playing for Ireland at Murrayfield. Then my cousins, James and Maitland Mackie, as well as my mother's cousin, Gordon Stephen, all played for the First XV.

And these great men conspired to get me to play rugby because as a youth of over fourteen stones and six feet tall, I was not the sort of player the university could pick up every day, even if I had not been a sprinter for the university who had even won a heat at the Great Braemar Gathering.

One of their principal pieces of bait was an institution called the rugby 'smoker'. This was, perhaps it still is, an all-male meeting in a soundproof bar in the Students' Union where the rugby club met to get drunk and sing the dirtiest and most politically incorrect songs you can imagine. For one whole season I thought these songs, many of which were witty as well as the filthiest male chauvinist piggery imaginable, were wonderful. But by the time I was well into my second year I understood why cousin Maitland had needed a couple of Carlsberg Special Brews before he could even attend, and I was so embarrassed by them that I used to hide when, after a game, the lads who were old enough to have had something better to do, burst into the same old nonsense about giant sex organs, tiny consciences, bestiality and abuse of women.

However, for a seventeen-year-old it was all very seductive, and so, when I broke down in a game for Rosemount with what is now well understood as 'groin strain' and was told by the doctors that it was caused by kicking a football and I should stop, despite the fact that a Rangers scout was at the game and he was thought to be watching the lad from Methlick, I saw my chance. I allowed myself to be invalided out of football and onto the rugby field. It sounds crazy, and even though you don't kick a rugby ball as much as a round football, it was to prove a very bad move.

As usual, beginner's luck was with me. My very first game was for Aberdeenshire 2nd Team against Dundee High School seconds. They put me on the wing because as a sprinter I must be fast, and as captain of the basketball team I must be able to catch the ball, but as I knew nothing about the game the farthest they could keep me from the ball would be best. Well, somehow the ball did get to me at least twice. The first time, I was able to run around the only opposition player in front of me and score. But the second time was much more remarkable, as I came to realise. I never even tried it again in my two years of playing. I found myself with a little bit of room but with no clear route to goal and opposition advancing on me from all directions. As I was about to disappear below a tide of tacklers I dropped the ball and let fly at it with my left and much weaker foot. I didn't even know enough about the game to be surprised when it sailed between the posts. Wing three-quarters who know what they are doing don't do that.

When the university reconvened in October I joined the old men of the university and went straight into the first team as a wing three-quarter. It was a ridiculous situation, but the University of Aberdeen's first game of each season was against the North of Scotland's select team, which would play against the Midlands to choose a North Midlands team. That team then played against the other districts to contest for places in the final trials for Scotland's team. A few of the beaten students would be picked for the North team so I can say, without telling anything more than a white lie, that my first senior game of rugby, and only my third of any kind, was a preliminary trial for Scotland. I only remember two things about the game, neither of which would have enhanced my claims to a game for the North. The first was when I found myself with the ball at my feet. I was told later that what I should have done was to tap it with my foot, then pick it up and pass it to our outside half, who would have kicked it into touch, gaining us twenty yards. But instinct developed by seventeen years dominated by playing football came out. I made to belt the ball up-field. I saw the opposition lose interest and withheld my kick. I had sold them the perfect dummy and was able to steady myself and gain the twenty yards by kicking the ball into touch myself.

Then there was the only time I can remember getting the ball in hand. I was in space and at speed. But my way to fame was blocked by the most ferocious impedimenta the North of Scotland Rugby Football Union could at that time field. Hugh Thomas was nineteen stones of second-row menace and Ron Comber the heaviest winger then playing, and both were final trialists for Scotland. The third was Gat Watson, a fellow farmer's boy, again a very rugged individual, who had played for the university a couple of years before. Now, none would have had any hope against me on the turn and all I had to do to make a name for myself was to chip the ball over them. If I hadn't managed to catch it I would still have had a chance. Had I got a lucky bounce I would have been in for a try. But I didn't even know I could kick it. I kept my head up and ran slap into their ambush. I don't know who got which bit, but one had me by the head, one almost cut me in half and the third got the legs. When I eventually got up I saw stars for several minutes. I wish I had done the obvious thing then and given up rugby.

However, there was no chance of that. As a first-team rugby player, however ropy, I was now friendly with the great men of the university, and everyone knew who 'that cocky bugger' was. When you are seventeen the old men of twenty-two are impressive, and I was impressed. Brian and Kenny Grassick were 'chronic' medical students. They did become doctors, but they took their time. We three had brought off a great coup when I was beer convener. We sold the most raffle tickets, which carried rewards totalling £85, with which we held a giant scrum in the basement of my mother's St Nicholas School in Albyn Place, surely the best-ever Aberdeen address for a student party.

And I became pally with Archibald Macdonald MacSporran a very intelligent West Coaster who had come up to do Divinity with Philosophy, but the Philosophy, helped by the women and drink, had won. He had been a great rugby player but never played again after Hugh Thomas crunched him in a tackle. But my great friend was Johnny Moffett. He was and is a delightful character from Northern Ireland. He eventually played twice for Ireland, including kicking the Emerald Isle to victory over England at Lansdown Road, but he then had an argument with the selectors and left in a huff for New Zealand with ambition to play for the All Blacks and beat Ireland. He had to settle for growing fruit and making his fortune and a fine family, who were astonished when I told them at the turn of the century what a star their father had been. His wife said wide-eyed, 'And there was I thinking I had just got this funny little Irishman no one else wanted.' Well, if she ever gets fed up of him I am sure there is still a queue of girls waiting for Johnny in Aberdeen. Moffett hardly studied forestry at all, but played a lot of rugby, golf, squash and cricket, and he drove quite the most dilapidated soft-topped sports car that had once been red. This thing (an Avon Special) had a top speed of about 35 miles an hour on the flat and yet when the police tried to stop him once for taking an illegal short-cut he took them on. It was down Mid-Stocket Road mind, quite steep, so he might have managed to break the speed limit at least. And when he got to the bottom he did a handbrake turn round the 'keep left' sign and tried to get away up the hill again. This car was notorious in Aberdeen for its noise, its fumes and for the bits that fell off it. One of the many girls who got a lift in the Avon Special actually fell out; Johnny turned up University Avenue off King Street talking fifteen to the dozen as usual, but when he looked to see if Isobel had appreciated his latest joke, she wasn't there. She was trotting along behind, unhurt. So Moffett, with his shock of naturally spiked blond hair, and his car, were known to every policeman north of Perth. And yet when the chase up King's Gate was given up he said, 'I'm sorry officer. I'm a stranger in toyne and I didn't know you weren't supposed to do that.' They let him off, which was no great surprise, as it was Moffett and those were gentler days.

Moffett adopted me and tried to teach me rugby. I don't think he succeeded although I was a natural ball player and I did get fitted in. We practised hard at a move which was very successful, very clever and so obvious it fairly showed what a lot of dumdums played the game in those days. It was universal practice that the backs of the attacking side lined up in a forty-five-degree angle to the gain line behind the scrum or lineout. When they won the ball it was passed from the scrum-half to the out-half, who passed it to the inside centre, who passed it to the outside centre. By the time it got to me on the wing, I had run at least forty yards and yet I hadn't gained any ground. Moffett's plan was very simple. Instead of all this passing between buttery fingers that would likely drop it anyway, instead of being forty yards behind the ball in the orthodox position, I would creep up almost level with the action. Moffett would then kick the ball as soon as he got it just far enough in front of me that I would catch it and be away.

We didn't use our trick all the time, but when we did it always worked. I can remember two times in detail. The first time was when we played against the Grammar School FPs (former pupils), who had a very strong team and had for years been the champions of Scotland north of the Tay. I don't know what gave him such confidence that we could beat this much better club who had no clueless converted footballers playing for them, but Moffett persuaded me to put all the money I had on the university to beat the FPs and there were plenty of takers at as much as three to one. We won the toss and booked the blustery wind in the first half. Moffett soon kicked a prodigious drop goal and then two penalties. Then with skilful use of the wind he manoeuvered us up to the Grammar twenty-five once more, and we played our masterstroke. With the FPs expecting another drop goal by the students who hadn't the muscle power to score a try he kicked the ball over their heads to me. My basketball skills chipped in and I caught it at full stretch and fell over the line. That was 12–0, and though they pounded our line all the second half, we ran out winners and I had over £13 of winnings to collect.

So, champions of the north of Scotland, we were off to London for our annual tour – down and back by coach for four matches and far too much beer. On this occasion our star match was against the Metropolitan Police, who were a first-class side. No doubt their midweek team had not been their strongest but we were given no chance of a result. If we were to win, Moffett's move had to come off. With five minutes to go we were three points adrift and a converted try would have done it. I sneaked forward, Moffett kicked high and just right. I caught it at full tilt and in the clear. Their line was guarded only by a thin line of defenders. But I had not learned how to run at defenders with the ball. Had I put my head down and charged there was no way they could have stopped me. But I was a very orthodox upright sprinter and that was how I ran at them. The defenders got hold of me, and although I got maybe ten yards over the line I couldn't get the ball down. I had blown it. We had a most commendable result in running the Met Police so close – and, amazingly, we were not finished. The coppers pressed hard and were threatening to underline their superiority when we won a scrum on our own line. Instead of kicking for touch, as was the only reasonable thing to do, our outside-half, Ian Macmillan (who became a lawyer in Canada), decided to make a break for it. He advanced up the open side, but, finding his way quite blocked, turned and danced back over our line and round the scrum to the blind side. Now, Mac was a slim man, quite fast, who jinked about in his running like a demented ferret. It was a sort of gallop from side to side: one–two, one–two. And he just kept jinking and running until he had got right past their astonished defence and only had about sixty yards to run, with half Her Majesty's Constabulary after him. Two of the opposition did catch up but failed to finger him. With a couple of last jinks he evaded them and fell, exhausted, over the line. Moffett missed the conversion but we had a draw, to that date the best result ever achieved by the Aberdeen Students on their Easter tours.

Mind you, the next year in London we pulled off a result which has been celebrated ever since. There is a dinner planned for 2010 to mark the fiftieth anniversary of our grand larceny. The opposition was London Scottish, one of the best teams in the country. They were fed up of this midweek fixture against a team of drunken no-hopers and they put out their strongest team to win by such a score that they would not be asked to play us again. All seven backs had played in international trials and they had an all-international three-quarter line of DWC Smith, PW Black, RH Thomson and IHP Laughlan. The half backs were AA

Waddell an All Black trialist called Blake. The full back was an Aberdonian and a final trialist, Eric Cruickshank. The pack was not quite as formidable, but it was not being led by a nineteen-year-old convert from football. The extreme shortage of big men had by this time led me to be moved into the scrum to play second row, where my inability to stop speaking and appealing to the ref led to my promotion to pack leader. As he couldn't stop me yapping and protesting 'aww Ref!', Toby Chalmers, the forward-thinking captain (perhaps the only first-class rugby captain proudly to wear a CND badge, he later ran the Northern Ireland Department of Agriculture), thought he might as well give me the right. It was one of those days which make sport such fun. A day when a team which has no right to be on the same park as their opposition wins.

Well, it was a story of desperate defence from start to finish, with flashes of never-to-be-repeated brilliance. Brian Grassick, our open side wing forward and still not Dr Grassick, and our sole player with any sort of representative honour, having played for the Scottish Universities team, ran riot among those internationals. He knocked them down and got up and knocked the one he had passed to down . . . time and time again and tirelessly. He must surely have been offside. Donald Mackay, who got a first in Economics and is now Sir Donald, in recognition of his role in Scottish Enterprise, never dropped a ball all day, and though he didn't make as much ground as the pack leader would have liked, he never missed a touch. Our hooker, an engineering student who became a company doctor, did his bit when it seemed that we must finally let those great men score a try. They had a scrum on our line; all they had to do was hook it and they were in beneath the posts. But somehow John Munro, who had very long legs for a hooker, managed to steal the ball and we won free again. My basketball abilities helped me win more ball from the lineout than we had dreamt of. That wasn't very much, but it helped.

Anyway, when we got to half time, 3–0 in the lead, we had already got far more than we had expected. Andy Gordon, an agricultural student, a left winger with not much speed but a lot of determination and a low centre of gravity, ran round his opposite number, Doug Smith – an international himself and a future British Lions tour manager. But the *pièce de résistance* was the ridiculous move with which we started the second half. It was Macmillan again. We all lined up to fight for his kick-off, which would be high to the left as was just about universal. But Macmillan tapped it straight ahead. The bounce favoured our only classy runner, George Masson, who picked it up, ghosted another ten yards and then chipped it forward. It was caught by Macmillan, who charged over without a hand being put upon him. In the absence of Johnny Moffett, who was now Junior Forest Officer Moffett, I missed both kicks at goal, but after thirty-nine more minutes of desperate defence we had the famous victory. Let us not try to describe the excess that followed that night in London, but we have been celebrating ever since.

Apart from being beaten finalists in the Elgin seven-a-side tournament, in which I kicked enough points to put my tally up to 100 for the year, that was about it. My rugby career was all downhill from there.

In the first game of the next season I caught the ball in a lineout and, putting it between my feet, prepared to push the opposition back. The plan was that my mates would give me a hand with the pushing. But our blind side-wing forward didn't follow the plan. He came charging round the forming ruck and dived on the ball right through my leg. My knee was bent from outside to in, my cruciate ligament was shattered and my cartilage mangled. Months later I tried again. In the first action I did a little dummy with no one near me and my knee crumpled. I was down and out. The blind side-wing forward said without a smile, 'I'm glad I wasn't anywhere near you this time.' It was a year before I had my operation, but it wasn't a success. The specialist was a Mr Hay, who passed me on to his registrar because he needed the experience. I don't know if that was why, but it never healed totally. It was fifteen years before I got free of the constant stouning of that knee. Team games were out for me, although I did manage ten years later to play football for Cambria in the third division of the Scottish Amateur League for a couple of years. And I did well in Highland

Games, but only by cutting my cloth. I putt the shot with an extra step instead of a hop. I high-jumped off my left leg, and made other adjustments to my throwing style, none of which gave me extra distance. At my best as a caber tosser, I did a 'centre of gravity test', which put a figure on the damage. I could do a vertical jump off my damaged leg of four inches, but off my good one twenty-four inches. I like to think of the cost of playing rugby as a five-sixths loss of athleticism in my right leg and fifteen years of sound sleep foregone.

If I sound bitter about rugby, it's because I am.

CHAPTER FIVE

A Serious but Susceptible Student

1960–62

The university brought me into conflict with my interest in Scottish Highland Games. All sport at the university in those days was amateur, although golfers and footballers mixed professionals and amateurs with no apparent difficulty. The best university golfers got to play with John Panton and Eric Brown in Open Tournaments, and the footballers played against the pros of the Highland League in the Scottish and Aberdeenshire cups. But in rugby and athletics it was a matter of immediate suspension for an amateur, not only to accept money for sport, but even to compete against anyone who had done so. This was a serious issue for me when I went up to the university, because I wanted to join in all the activities and I had already won quite a bit at the Highland Games and in the agricultural shows. As early as my fifteenth year I had won £4 at New Deer Show, and that was thirty-two times my weekly pocket money of half a crown, so it was serious money. Luckily, I had taken precautions against being banned from amateur sport. After my triumph at New Deer I went to the Ellon Agricultural Show, where they had serious prizes. There were four events I could enter, promising £4 each for first, £3 for second and £2 for third. Ellon also had serious bureaucracy. Whereas at New Deer anyone who appeared at the start was deemed to have entered, and the first three to the finish got the prize envelopes, in Ellon I had to present myself for registration at the secretary's tent. When the secretary asked my name it suddenly came to me that if I entered as Charlie Allan my chance of fame as an Olympic athlete would be gone forever. Now, I had been the night before to Pittodrie to watch Aberdeen reserves play against some continental visitors and our centre-forward, who scored the winner, was called Ivor Smith. So that was it. Without hesitation I gave my name as Ivor Smith and competed under false pretences for about six years. This was quite uncomfortable for my wife, who had to be Mrs Smith on Saturdays in the summer-time.

The subterfuge was not important for the Olympic movement, but it did allow me to compete at the university. Some of the true blues didn't want the hallowed turf of King's College sullied and would have banned me. But they couldn't prove that I had been to the Games. Things became delightfully ridiculous when I won the silver medal at the great Oxton Games in Berwickshire. It was reported in the *Daily Express* that an Aberdeen University student had travelled all the way down to Oxton and won the high jump and all the heavies. But there was no photograph and the student's name was Ivor Smith, who told the *Express* reporter, 'The prizes weren't much, but it was the medal I came down for.' Still, the forces of virtue were determined that I should not be allowed to compete at the annual university sports. When I turned up in good time for the shot putt I saw what I took to be a reception committee waiting at the gate for me. There then followed a piece of cloak-and-dagger subterfuge which surely suggested I should consider a job in Her Majesty's Secret Service. My kit was in the pavilion and my road to that was heavily guarded, so here was what double agent Ivor Smith did: he went to Peter Craig-myle's sports shop half a mile away up King Street. There he tried on a pair of running shorts. They fitted, and Smith said he would take them – in fact he would just keep them on. Then it was back to King's, where he hung about outside the ground, doing warm-up exercises in the street until he saw through the railings that they were ready to start the shot putt. The judges were surprised when it was his turn when Charlie Allan climbed over the wall (seven feet high), apologised for keeping them, picked up the shot and threw it. The reception committee were on the other side of the ground and by the time they saw me they were too late. If they had intervened after my first throw they could have banned me all right, but everyone else in the shot putt competition might have been banned as well for competing against a professional. It was that ridiculous in those days . . . at least, I was able to convince them that it was.

Incidentally, that victory over authority was important in a way that I could hardly have anticipated. It was in 1960 that I turned up at King's College for the first training session of the athletics season. I remember it so well. I came round the corner of the pavilion and saw a jolly group of girls trotting round the track in what turned out to be the trials for the team which would contest the 440-yard race in the inter-university meetings throughout the season. And way in front of the pack was a pair of very neat sky-blue shorts simply flowing round with huge and elegant strides, below a fine pair of shoulders and medium-long fair hair. I said to Odelle MacKinnon, the staff member in charge of women's sports, 'She looks OK.'

Miss Mackinnon, who thought I was speaking about her running style, said with a snort, 'Huh, that's Fiona Vine! She held the 440 record for St Andrews. She won everything last year.' She was now a junior lecturer at Aberdeen and was eligible for the sports teams because she was registered for a PhD. She turned out to be very good fun, as well as having that very good job, a car and a flat. Though she had several boyfriends, there was nothing which I felt was beyond challenge. We might have met anyway, but if they had succeeded in banning me from the athletics team I fear I might have missed what has turned out to be a most agreeable boat.

Having introduced Fiona, I can't really go much further without telling you that one thing led to another . . . and another. Now I am, I think, an unusual person in that, unlike any stereotype I have seen of young men, I always looked eagerly forward to the day when I would have a wife and be married. When I was about five I said in a desperate little voice to my mother, 'Oh Mummy, I would marry you but you'll be too old to run for the bus.' At my boarding school in Devon whenever I had a girlfriend, which was most of the time, as I was a susceptible youth and not unattractive, I used to count the years till I could be married. I remember thinking that with my education to get in and a degree to get I would be bound to be twenty before it was possible, but that seemed ages away. My desperation to be married and get on with the rest of my life was one of the reasons that I persuaded my teachers to let me, after a very undistinguished performance in Ordinary level, sit my Advanced GCE exams in a year instead of the usual two. So the blue running shorts provoked another outbreak of dreaming.

And the dream came true in less than a year, and in circumstances which were not exactly planned . . . not consciously at any rate. We got engaged, which I thought a poor substitute for marriage. A diamond like the ones that were beginning to flash on the wedding fingers of the girlfriends of my pals at the university was out of the question, but Fiona accepted my very nice ring of Scots agate set in silver. That extravagance, of over £20, was made possible by my first-ever all-night game of poker with my friends John Adam and Sandy Henderson (the very well-known Aberdeenshire farmer who had been a loon at Little Ythsie when I was being a refugee and staying with my granny at North Ythsie).

My mother made it clear that she was delighted but that we would wait till after I had graduated. I really don't know what mixture of incompetence and desire to be married was responsible, but at any rate within the year we were pregnant. For ourselves, we were delighted; we had a house, a car, one job and the partner we wanted. But there were two other lots of people involved . . . we needed the blessing of our parents and my grandparents.

We went to Broughty Ferry first to see Fiona's widowed mother. She put a brave face on her displeasure and let it be known later that she had been impressed that, having told her the news, I retired to get on with an essay I had to write for a class the next day. When we went to Little Ardo to tell my parents my father gave Fiona the biggest hug of welcome you can imagine. There seemed a hint of desperation in it. Perhaps he was remembering the circumstances of his own birth, when his mother dumped him in an institution and fled to Canada. My mother called me a 'silly ass' and lost no time in planning the wedding.

Then there were the grandparents who looked after me during the war. I went in fear to North Ythsie and found them in the drawing-room where I had played cowboys and Indians around the giant furniture during the war. Granny was knitting on the left-hand side of the fire and the old man by his desk at the right, handy for the phone by which he did most of his farming and the store of envelopes on the backs of which he did his cashflow projections, calculations of profit or loss and notes of things he would tell people the next time they phoned him. They were a saintly couple whom I worshipped just this side of idolatry. I told them our news: that we were to be married and that we had all we needed to look after ourselves and a baby. Remember, they were Victorians and pillars of the Church and the Conservative party. They didn't say much, but while I was pleased, indeed excited by the turn of events from my own point of view, I thought I could see it from theirs, and that I had let their expectations down with a terrible dunt. I could have stood a dressing-down, but what I got was a severe silence. I think the first time there had been anything other between us than happy chatter. I was overcome. I cried, huge sobs of shame at causing them disappointment. But I left the wonderful drawing-room at North Ythsie embraced by their warmth. The Old Man cemented that warmth with a cheque for £100 delivered to my parents as a wedding present for us 'because there will be crises'. Maitland Mackie was seldom wrong, but he was in our case. And the Old Lady was quicker to acknowledge that. When we bought our second buy-to-let flat in Aberdeen not much later, she said in a most reassured way, 'Your marriage is going to be all right.' And it was the Aberdeenshire 'aa richt', which means somewhere between good and very good. When she said that I nearly cried again. There were no crises. We always managed comfortably by cutting our cloth, but I have always cherished old Maitland's generosity, for £100 was roughly five times his usual present.

After fifty years I think I misunderstood the old people's reaction. After all, Mary Yull's illegitimate sister Isie had arrived at Little Ardo when she was about ten, on the top of a horsedrawn cart with all her possessions. And Maitland Mackie's reaction may have been so subdued because he was remembering the day in the 1930s when his eponymous son told him that he had got a lassie pregnant and old Maitland covered it up so successfully that the truth only came out after he, his son and even the baby, were all dead. My grandparents may have been Victorians and I think they were disappointed, but they cannot really have been shocked or unduly surprised.

When I said Miss Vine had a flat I wasn't giving the true picture. Just outside Aberdeen, about a mile from Bridge of Dee along the south Deeside road, is Banchory-Devenick Estate. Now, as you would expect, the laird had a dower-house to keep his mother in a style suitable to the former mistress of a big hoose. And Fiona had got the tenancy of a wing of this dower-house, set in the beautiful mature woods of Lower Deeside. Drumduan was a perfect setting, so we were married there. The powers that were (my very nice but bossy mother) thought a minimal wedding was appropriate. So we just had the three parents and the minister. It was a beautiful early spring day and the bride and groom warmed up for their wedding with a walk through the woods and fed the ducks on the artificial lake. I remember thinking how easy it was going to be to get one for the pot when the laird wasn't looking. Then we went in and stood before our own fireside while the minister said the old words to the select company. I only remember one detail. When he had to say, 'With all my worldly goods I thee endow,' the student's father gave a little chuckle. No doubt from where he stood all the giving was by the lecturer with the house, the car and the bank-book. Then we had champagne and off to Raemoir Hotel for a meal of our careful and indulgent choice. With only five to feed, as the minister couldn't stay, we could damn the expense. We had the happy couple's specially designed menu. Field mushroom soup with big chunks of the fungi in it. Lobster thermidor, roast pheasant and trifle followed by crumbly cheese biscuits like those my granny produced every morning at ten for the coffee, and ripe Camembert. I don't remember the drinks side of the menu but there was plenty and it was good. It was a lovely occasion and none the worse for us being so few. I have always thought, and had it confirmed time and again since, that the important thing about such occasions is not the wedding anyway, but the marriage, and this marriage has been a blessing indeed.

Mind you, my cousins and friends on my mother's side may have thought they were done out of a wedding and they were not for that. One night soon after, a knock on our door at the dower-house turned out to be about twenty young bloods who carried us off to the Athenaeum, the best restaurant in Aberdeen, where they had booked a private dining-room in which they gave us a splendid meal. It was good of them.

Now everything was in place for a complete change in the student's lifestyle. Rugby, with its competitive drinking, was finished by injury, and we had settled who I would marry, so there was no use in all that looking for partners or standing about late at night in draughty doorways. The way was now clear for the playboy to get serious about building a life, and the necessity to do so was gladly embraced. It was reinforced when Sarah was born and her mother retired from her job. Now it was clearly down to me to get a good degree and a job in which to use it.

CHAPTER SIX

A Don at Last

1962–64

The other subjects I had gone up to Aberdeen to study had eliminated themselves early on. Mainly that was down to what I thought of the people who taught them. When the distinguished poet, Professor Claude Colleer Abbott, who was a friend of my parents, heard I was to study English Literature under Professor Duthie at Aberdeen he said 'His little book on Shakespeare I found quite appalling.' I soon saw why. This dry stick of a man could make a bore out of anything. To me Shakespeare meant ideas and excitement, comedy and poetry, and exemplary use of the English language. But Duthie's greatest delight in the Bard of Stratford was something called 'textual criticism'. If he could find more prepositions in the second half of a play than in the first half he was made up. That kind of English was clearly no use. And Philosophy, while wonderful no doubt, took a year longer and with a wife and baby to think about I was running out of years. Besides, I realised that even though Plato may have been right that the apple-tree could not be proved to be in the garden even if Theaetetus wasn't looking at it, worrying about that might not make a financially rewarding career. I thought I would carry History on as far as the ordinary class in Economic History, where I won my only university class prize. But Economics seemed to be the way forward, and there were three reasons:

(1) There was a demand for people with Economics degrees;
(2) I could understand what it was about; and

(3) Because the staff at Aberdeen University's department of Political Economy were such outstanding teachers.

The roll of honour was Professor Henry Hamilton, Annie MacDonald, Ken Alexander, Roy Houghton, Malcolm Gray, Harry Richardson and Philip Whelen, and they were really interested in us. We felt free to ask them about anything, and at any reasonable time. We saw them in class, we went for our morning coffee with them, we lunched with them. In and out of class we debated the great political and economic questions of the day as well as whether Aberdeen FC would ever win anything. An idea of the easy relationships they had with their senior students can be had from the following anecdote, with which Sir Kenneth Alexander (later professor at Strathclyde, Vice-Chancellor at Stirling and Chancellor at Aberdeen), in the days when he was just Ken, started off one of his classes in Industrial Relations: he had had to take his wife Angela into hospital the previous night when she was suddenly taken ill. He had to tell their five children, who were watching telly: 'Now, children, you've not to worry. Mummy's going to be fine, but I've just got to take her into hospital and she'll soon be home again, so you've not to worry.' As he hurried off they called after him, 'Bring back chips.' It was a good way to start a lecture and we were indeed privileged to have such a relationship with our teachers. Those academics created such a congenial atmosphere that we honours students couldn't imagine anything better than to stay on in the universities and enjoy all the freedoms we had enjoyed for four years. An academic job promised a good salary (the starting salary was better than what top industrial giants like Shell and Unilever paid new graduates). It was also a secure job, from which you couldn't get the sack unless you did something called moral turpitude, and even then you'd have to do an awful lot of it, leading, even if you had to wait for dead men's shoes, to a very good pension. There were four of us who were reckoned to be capable of getting a good enough degree to be offered a life of ease, and I decided to get off my mark by applying for the first rung of the academic ladder. That was a research assistant's job at the University of Wales in Swansea.

I had heard of the professor there, E. Victor Morgan, a renowned economist who had done work which many thought important, on money, and I reckoned that at a salary of £500 a year the competition for the job wouldn't be too hot.

Telling you how I got on in my interview is an act of great intellectual honesty, and fifty years later I can only tell you through my pain. I could not have made a worse job of presenting my credentials for the job of research assistant in International Trade Studies. I was awful. Luckily I can only remember part of the disaster, but it went like this: after preliminaries about whether I had had a pleasant ride down and whether the train was late, Professor E. Victor Morgan asked me why I had chosen to study International Trade as a postgraduate subject. Well, I should have said that I had a wife and child and I needed a job, and International Trade was the job description on offer. But no: I had to try to convince the panel of learned professors that I found trade fascinating and it was such an important subject. The next question was my downfall. In my four years of studying Economics, which books had I read on International Trade? I knew quite a lot of the theory of International Trade from my lectures and from text-books, but I couldn't give them one book, nor even one author of any work, in the field of International Economics. If the floor had opened I would have dived quickly and with such relief.

I did not get the job. It was a much sadder young man who took the train back to Aberdeen.

But the young man was also much the wiser of his terrible humiliation in Swansea. A much better job came on offer at the University of Glasgow. It offered a salary of £900 a year and, with the extra £50 they offered their lecturers for each child, I would have almost twice the money I had failed to get in Wales, if I got this job.

If I made a terrible hash of my interview in Swansea, and I did, my performance in the Glasgow interview was the stuff of a master class. First, I found out what the job was about. For that I phoned the old friend who used to come for his summer holidays at Little Ardo. Donald Dewar had finished his first degree at Glasgow

University and was now studying Law. He had friends in the relevant department and was able to tell me that there were in fact two jobs. One was Assistant in Business History and one was Assistant in Urban History. He also told me that there were a large number of candidates, many of whom had already finished their doctorates, their second degrees, while I was only waiting to sit the finals for my first degree. It was clear that I had nothing much to lose and I was determined to give it a better shot than I had done at Swansea. I also found out that the professor, Sydney Checkland, a delightful Canadian who still carried a very distinguished limp from a wound received as a pilot in the Second World War, was a bit of a maverick. He might use different criteria in choosing his staff than others. It might do no harm to take a punt on standing out in the crowd of applicants.

By the time the interview was well under way and I had already dealt with the tricky question of what sort of a trip I had had and what the weather was like in Aberdeen, I was ready to show off how well-read I was. I was asked whether I wanted to study Business History or Urban History. I didn't answer straight out, but asked if Lewis Mumford's *The Culture of Cities* was the sort of background reading that would be helpful in Urban History. Then, knowing of the professor's Canadian origins, I managed to slip in reference to the American 'Middletown' studies. I don't suppose there were ten students in Britain, including architects and town planners, who had read both of those classics, so I had definitely got myself on the shortlist by then. But I'm sure I got the job because of my answers to the next two questions. 'Why have you chosen to apply for a job when you haven't got a degree yet, instead of studying for three years and getting a PhD first?' Remembering the fool I had made of myself in Swansea, I wasn't going to say that I had a burning desire to study Urban History. I told them that I had got my wife with child and that I needed a good job to keep us. Then the coup de grâce. 'And Mr Allan, where do you see all this leading you?' asked Professor Checkland. Without a moment's hesitation the twenty-two-year-old undergraduate replied, 'I aim to be Dean of the Faculty at the

Massachusetts Institute of Science and Technology.' They laughed and laughed. This boy was aiming high.

I got the job. Professor Checkland took me on without qualifications in preference to several candidates who had doctorates and two dozen or so who had already got the good honours degree to which I aspired. I did my bit by getting, three months later, the first-class honours he had taken on trust, but I have always been grateful to him and admired his willingness to take that chance. Sydney Checkland was no stuffed shirt.

I have always made a bit of fun of my thirteen years as a lecturer. I have often said that a university lectureship was more of a salary than a job. Ten hours a week was considered a big lecturing load and the year could include twenty weeks or so when there were no students around, so you could do what you liked. Some of my colleagues did treat it as a cushy number. Some just trotted out the lectures they had been given when they were students, with little or no amendment. There was marking to be done, of course, and you were expected to be available to students to explain out of class what you had failed to explain in class. A very few dealt with that by not being there and others by making the students so uncomfortable about their failure to understand that they didn't ask again. I liked the students and always did my best to make them feel at ease by sharing my ignorance with them and doing my best to show them where they could find out more, even if I couldn't tell them the answers. I was never so proud of myself as when I heard my boss at Strathclyde replying to the question, 'How is Allan getting on with the students?', 'Well, there's always a queue at his door.' Then there was research. As well as teaching, academics were supposed to add to the sum total of human knowledge. You were supposed to publish articles in learned journals and even books. That was how to get on in academic life. You could be as bad a teacher as you liked. You could treat the job as a salary rather than a task as long as you got stuff in the academic journals. Some of my colleagues just accepted that they were on a salary scale that rose gently each year and from which you could only be sacked for moral turpitude, and did no research at all. But that wasn't an option for Young Allan. If I were to get to be Dean of anywhere I had to get going. So while university life gave me a lot of freedom to make my own choices, one of the choices I made was to work pretty hard.

First I had to complete my Master of Arts in Political Economy at the University of Aberdeen. Well, the cunning I had displayed in my interview for the job in Glasgow also stood me in good stead in my studies. I decided at the beginning of my last year that I had no hope of learning everything I was supposed to know. I also studied the past exam papers, where I noticed that certain questions appeared regularly in various guises. I thought that if I chose to study half the syllabus only, I would end up with enough questions to get by. So I left almost half the curriculum out altogether and concentrated all my efforts on the half I understood best. The exams consisted of seven papers, each with perhaps ten questions, of which you had to write for three hours on four. My cunning plan worked well on two counts. First, I always had at least five questions I could handle, and second, as I had not studied half of them, I didn't waste precious time choosing which four to answer.

Professor Alex Kemp, now the international oil expert who's always on the telly telling us why things are going to get even worse before they get much worse, had won the class prize every year and he was expected to get the gold medal. But David Forsyth got the top spot and both got first-class honours. That was to no one's surprise. But there was some astonishment when Allan's name also appeared on the list of *Primi Honoris*. I can't say that I was surprised – though perhaps I should have been. My main feeling was that I had got away with it. I had applied for a job way above my status, but now I had given room for hope that the status would be only delayed. I had got my foot on a very secure ladder which removed so much uncertainty from my life. I knew that I would always be able to support my family.

It was also a great day for Fiona. She had worried that people would be thinking that by marrying me and making me a father before what many would have thought was my time, she had spoiled my chances of academic success. In fact, without the settled status she gave me I would never have got a first. The truth of the matter is that it was her academic career that suffered from our association. When we met she was already on the ladder to which I aspired, as a junior lecturer in the Department of Clinical Chemistry at Aberdeen University. She was well through her studies for a doctorate; indeed she had done all the research and had had an article about it published in a top academic journal. But serial motherhood, and looking after a house with me in it, beat her when it came to writing up her research. And yet she worried that she had jeopardised my career.

There was plenty for the new junior lecturer to worry about in taking up his new job. He had never given a lecture on anything in his life. He had never once had the courage to speak in the university debates at Aberdeen, although he attended them every week. Indeed, the only speech of any kind he had made was the vote of thanks on behalf of the guests at the Fyvie Youth Club – Girls' Club Christmas Party when he was seventeen. I remember it word for word, 'I would like to thank Mrs Cruickshank and the girls for a great party and I'm sure you would too.' That took all of six seconds and it had taken a long and nervous time to compose. But a university lecture had to last a minimum of fifty minutes. How was a guy who had only studied half the syllabus of an Economics degree going to manage that when this job was in Economic History? But you wouldn't have thought I would be nervous about the first day in my new job. No teaching was to be involved, I was just there to meet my new colleagues, find my office and my desk and arrange meetings with Professor Checkland to discuss what exactly I might do for a research project.

I had been so nervous the night before that I had walked and timed the distance from the flat in Lacrosse Terrace, down Hamilton Drive, across Great Western Road and along Bank Street. Then up University Avenue past the Students' Union (the Men's Union as it was in those days when women were not allowed in and had their own Union). After sixteen minutes' brisk and businesslike walk, I arrived at University Gardens, the imposing Victorian terrace that had once provided homes for medium-wealthy Glaswegians but had since been taken over by the university. No. 21 housed the Department of Economic History. So I was confident I would be on time for a nine o'clock start, but not so early as to let them know how nervous I was.

Up I went at nine a.m. precisely, up the steps to the great Victorian doors at No. 21. In I went to be greeted . . . by nobody. There was no reception desk and it soon became clear that there would be no reception of any kind for some time. The family farm was 165 miles away and across a huge cultural divide. James Low and his men would already have been working for two hours.

Eventually I came across a very Glasgow lady who was clearly a cleaner. She seemed nervous of the nervous new member of staff. I later became quite friendly with her and she told me that she was supposed to be finished by eight o'clock, but in her ten years in the job I was the first academic she had seen before ten. As she liked a lie-in, she had adopted flexi-hours.

Just before ten o'clock, the distinguished figure of Professor Sydney Checkland came limping down University Avenue. And just after him came the senior lecturer, soon to be professor, Roy Campbell, who had taken me for coffee at the time of my interview and with whom I had had an argument about foxhunting (he was in favour) and council housing (he was against). Campbell showed me my room and my desk. Some time after eleven the others with whom I was to be sharing my room arrived: Peter Payne, later Professor of Economic History at Aberdeen University, his wife Enid, who had a job as his research assistant, and Bob Tyson, who was another new boy at the bottom of the ladder but who had already done a postgraduate degree in Manchester. Then there was a third new assistant, Clifford Davies, who had come from Oxford, and Dr John Kellett, both of whom were important enough to have rooms of their own.

Kellett, whose research interest was Urban History, was put in charge of guiding me in my new job. His method was to be pleasant to me and let me do what I wanted. He was an amusing man who liked my story of my interview at Swansea because he had had a terrible time looking for a job himself. After many disasters he had taken an evening class in positive thinking when that was a very new 'study'. He had hoped that would help him with his stammering and other terrors in interviews. So he sat outside the torture chamber saying to himself, 'This time I'm going to get the job. I'm going to get *this* job. I *am* going to get this job. *I am* going to get *this job* . . .' When his turn came in he went to face the vast panel of learned faces. The chairman said something to the effect, 'Ah, good afternoon, Dr Kellett. Did you have a good train journey up from Birmingham?' All poor Kellett could say was, '*I am* going to get *this job.*' Secure in the job he had at last got at Glasgow University, John could laugh at that experience. But he went really pale and quiet when he told me of another of his interviews. He had got off to not a bad start, he thought, in his quest for a Commonwealth education job, the interviews for which were being held at the London School of Economics. He had managed the teaser about how the train had been on his trip down from Birmingham and whether he had been able to get a seat on the Underground, always being careful to be polite and call the chairman a deferential 'sir'. Then they got onto his academic interests and he explained what his doctorate had been about and was making a pretty good job of explaining why it had not yet been published, when he began to notice a sort of atmosphere growing in the room.

Dr Kellett started his next answer by speaking directly to the Chair and looking for clues in his face as to what this atmosphere might be about. 'Well, you see, sir . . .' It dawned slowly and horrifically. The Chair wore a tweed suit and tie, and had a hairstyle which was in those days called an 'Eton crop'. It would almost have been a 'short-back-and-sides' if worn by a man. But this soft deep voice did not come from a man. Nowadays, when women frequently wear masculine styles and do jobs most often done by men, they might not object too much to being mistaken for a man – but this was over fifty years ago and John Kellet had made a bad mistake in calling the chair 'sir'. He did not even try to recover. He ran.

My supervisor and I became firm friends in a casual sort of way, meeting on neutral ground in pubs for the occasional pint, but never three. He did a little to guide me into Urban Economic History which, when I took the job, was little more than three words to me. Among the few memorable things he told me was that there were a number of ways of finding out what things were like in the towns of long ago, and one way was to ask old people. He had employed this method in researching employment in Glasgow and he got hold of the oldest Glaswegian it seemed that had ever worn grey hairs, not that he had many of those left . . . nor teeth. 'So what was the main economic activity on Sauchiehall Street in the old days?' Dr Kellett asked the Ancient Mariner. I don't suppose Kellet told me the old man's reply word for word and I have only remembered the gist of what he told me: 'Oh aye son. Well, hooers was aye big in Sauchiehall Street. I mind there was this place and she took ye up the stair and intae this big room and there was jist mattresses in the flaer, maybe a dozen o' them.' Kellett was shocked and elated. This was clearly a piece of important urban history. But of course, if you want to elevate hearsay to history, you need dates. 'So when would this be?' he asked. The old man screwed up his face with the effort of concentration, for he knew how important accuracy was to researchers from the university. 'I think it would be about June,' he said eventually. It was 1962.

When I got down to doing research of my own it was a far more mundane subject, but one perhaps more respectable. The City of Glagow led the world in the nineteenth century, not just in shipbuilding but in town planning, more accurately urban renewal. In 1866 they got an Act of Parliament setting up the City of Glasgow Improvements Trust, which set about buying up and knocking down the whole of medieval Glasgow, including its sixteenth-century university. Against Glasgow's heritage it was a crime, everything was obliterated bar two houses which still stand today. How they did it, how they got rid of all the worst slums and redeveloped the whole sixty-six acres of the old city centre, with broad streets capable of taking modern traffic, with fine shops offices and residences, and at no cost to the rates, had never been studied and written up. That was what the farm-boy from Aberdeenshire did for his first academic job. It was a far cry from stookin sheaves or hyowin neeps.

The records of the City of Glasgow Improvements Trust were kept on the ground floor of Glasgow's magnificent Victorian Town House in George Square. There I was given a desk, regular cups of tea and a sort of sceptical respect. 'He's only a laddie – what could he know?' Well, not very much, as it happened, but he did find out quite a bit. The records were mainly minutes of every meeting of the Trust. They detailed every purchase of slum property and every sale of sites once they had been cleared. In the course of a few months I read right through the forty years of those minutes and right through the *Glasgow Herald* of the period, where there were reports of all the thousands of disputes that had broken out, and the heads that had been broken, as well as the rest of life in the *Herald*'s area. I enjoyed it, and when I eventually got a long article accepted by the *Economic History Review* I felt it was some repayment for Professor Checkland's leap of faith in giving me the job without qualifications, and so, I think, did he.

So that was the research I did at Glasgow.

That part of my first real job was comfortable from the start. The same could not have been said of the teaching. The first class I took was a tutorial with maybe ten students. My job was to listen to a short spiel by one student and then to try to control a discussion among the students. That first class had a good topic. It was the writings of Thomas Malthus, an eighteenth-century economist who observed that, as with weeds in the garden or rabbits in the field, nature endowed mankind with a far bigger ability to breed than was ever needed to keep the population steady. Thus, however much you increased the amount of food available there would always be a huge proportion of the population on the breadline. If you doubled food production, very soon you would have double the number of people and double the number of people starving. That was the sort of thinking that got Economics the nickname of the 'Dismal Science'. And it was pretty dismal. The only hopes of keeping the population down to the food supply were war, disease and 'vice'.

Vice might reduce population pressure so that starvation might be reduced; no doubt there were other aspects of it like infanticide, late marriage and homosexuality, but the part of Malthus's vice that has stuck is birth control. The poor man even had a female contraceptive called after him – the Dutch Cap is sometimes called the Malthus Cap. At any rate, I was fairly sure I would manage to get some sort of discussion going with a group of eighteen-year-olds on the ideas of Thomas Malthus.

Well, the eighteen-year-olds were disappointing. They were very backward at coming forward which was, had I but realised it, what you had to expect from West Coast bairns who, having gone to a local school, had now gone to their local university and still lived at home. It may be a little different now, but compared with my contemporaries at Aberdeen they were very unsure of themselves. But the schoolchildren were not alone. In the class there were two mature students. That should have helped, but it didn't. These two men were in their thirties, ten years my senior and wore black cassocks. They were Roman Catholic priests. They had views on vice. The young assistant lecturer was embarrassed but he stuck to his task. He tried for fifty minutes to persuade these men of faith that, though they might be right that God would provide and that Malthus would rot in hell, in Social Science faith is no use. What this class dealt in was evidence. I don't remember seeing those men again, though I may have done.

The lecturing was a lot of work the way I did it, but it wasn't a problem. I only had a total of half a dozen or so lectures to do in my year at Glasgow. The students had to know about the whole of British economic history from the Industrial Revolution to the Second World War, so I just had to choose a bit I knew something about. I don't remember what it was but I did know enough to fill fifty minutes, six times. I didn't trust myself to just speak about my subject. Terrified that I would dry up, I wrote it all out in full, and I tell you, even reading slow, that is a whole lot of writing. I learned with time that it was far more than the students had any hope of remembering, but anyway, the new lecturer got by, and the students were able to study something else for their exams and no great harm was done.

So the job was all right. Professor Checkland was a most sociable man in a formal and upmarket sort of a way. He always had something to say when you bumped into him and it was usually interesting and often funny. He didn't take his staff down to the boozer and fill them full of drink or anything. He held receptions at his house where his staff were invited because he was a friendly fellow, but also I think to show what a lot of bright people he had in his department. His wife Olive was the only really accomplished hostess I have ever seen at work. Immediately any guest arrived she would greet them and say she really wanted them to meet . . . and off they'd be whisked. No one stood about wondering what to do at the Checklands' parties. That was particularly beneficial to the new assistants. At our first such occasion Mrs Checkland grabbed Fiona and me and rushed us over to two eminent members of the university establishment, saying, 'You must come and meet Professor Robertson and Professor Cairncross.' Those were famous men. Denis Robertson was in charge of planning the economic foundations of Livingston New Town and Alec Cairncross, then the top economic adviser to the Treasury, had written the textbook we had all used in our first-year Economics class at Aberdeen. They had obviously heard about the loon from Aberdeenshire's interview, for Cairncross asked, 'And *did* you get a first?' When I answered that I did Cairncross said, 'Oh, well done, and Aberdeen is a place where a first does still mean something.' That was a proud moment.

Apart from the Catholic priests I cannot say that I remember any of those few students whom I taught at Glasgow except for one very bright first-year student who may have been called J. Roy Hay. He gave me a little voluntary help with my research when it got going, but I lost all touch with him after that year. On the other hand I do remember some of the very talented and politically ambitious students who were at Glasgow University at the time. First was Donald Dewar, who had been such a help to me with my interview for the job. He lived with his parents in a huge house in

Royal Terrace. There he had a basement flat where he could entertain all his friends in comfort and out of the parents' way. I guess I was a bit of a curiosity as a most unacademic academic. It may have been Donald's revenge for the humiliations the Little Ardo loons and I had inflicted on him at football when he had visited the farm, but for whatever reason I was invited to Donald's Dive on a number of occasions. One evening I remember in particular. The company of perhaps six included: Jimmy (Lord) Gordon, who founded Radio Clyde; John Smith, who became leader of the Labour party but died young; and Neil MacCormick, a Scottish Nationalist stalwart who soon became Professor Mac-Cormick. There was also a young Tory who, Donald explained to me, was very bright, but, as he was a Catholic, had no chance whatsoever of getting anywhere in the Tory party. Donald was clearly nonplussed as to why his friend would not just join the Labour party. I was staggered that religious bigotry, which I never knew existed until I moved to Glasgow, could have such con-sequences in Scotland.

I was intrigued by those obviously very bright young men and intensely interested in politics. But they had no interest in, or at least I didn't ever hear them discuss, matters of political principle. Politics to them seemed to be wholly devoted to who would get the vacant candidature for such and such a safe seat and whether so and so should accept the nomination for a hopeless seat to win their spurs and then be promoted to a seat where there was more hope. These were quite unlike the political animals I had known in Aberdeen, who argued all the time over the class war, nationalisa-tion and nuclear disarmament. I didn't know what to make of it, but the only time they got anywhere near to discussing principles in my hearing was when Donald asked me about my politics. I wanted to tell them that in all issues I instinctively took the side of the workers, but that that didn't mean I was in favour of the Labour party sticking to Clause Four of their constitution, which encouraged further nationalisation of the commanding heights of industry. What I said was 'Emotionally, I'm a socialist.' This was greeted by hoots of incredulous laughter. It may not have been well put but the young teuchter knew what he meant and stuck to his guns. I wonder if it was refreshing for these young Turks to have to think a bit about what all their political manoeuvring was for.

Ming Campbell wasn't there that night, but I had met him on the athletics field when I was a student at Aberdeen and he was already studying at Glasgow. Neil MacCormick, who became an eminent Law Professor and Scottish Nationalist, was a very bright and also a very nice and unpushy man. And his cousin Donald MacCormick, the broadcaster, I met outside the dive. They were a bright lot, and they honed their skills in political debate in the Glasgow Union's Debater Chamber, as well as in discussions in the Dewar basement. Those debates were most impressive, and the impression was of national standing, for Glasgow students were frequently winners of something called the Observer Mace, a sort of UK national championship of glibness and point-scoring. I went to a few of those debates and sat and observed there in the gallery, as I had in the very similar chamber at Aberdeen. There were many good speeches and many good jokes mostly lost in the ether. There was one occasion when a visiting American stood up to talk; he was on a Fulbright scholarship and some kind of champion of speeches at Yale University. After a laudatory introduction he got up and, with lugubrious voice and gestures, tried to grab his audience. 'I come from Yale,' he intoned. 'Y for youth, A for ambition, L for leadership and E for en'erprise.' While the rest of us marvelled that anyone could be so crass, a very Glasgow voice called out, 'Thank Christ ye dinnae come fae the Massachusetts Institute of Science and Technology.' That is my favourite 'Tak doon' of all time, except that the Fulbright scholar wasn't taken down much as he hadn't a clue what everyone was laughing about. I wish I could say I was there at the time, but in all honesty I can't remember, though I have told the story so often I do believe I was indeed up in the gallery as usual, saying nothing.

We lived in a flat while I worked at Glasgow University. It was our first try at owner-occupancy and it was a pretty jammy start. I don't know where else in the world we could have got, for the money, so much house so graciously appointed as we got at 21

Lacrosse Terrace in Glasgow. It was a ground floor, own-door tenement flat with a very grand lounge with wonderful plaster mouldings, three double bedrooms, a tiny but adequate converted corner for a bathroom and a roomy kitchen with linoleum-covered worktops and the ricketiest shelves I had ever seen. Lacrosse Terrace is a quiet cul-de-sac on the banks of the River Kelvin, on which we had salmon-fishing rights. We got all that for £700 pounds, and not just for a month. For that money we got the title-deeds.

There had to be downsides. I could see some immediately. The salmon-fishing rights were definitely out of date, for it had been many years since salmon had been able to survive in the Kelvin. Indeed, our salmon river was really an open sewer with more wrecked prams in it than fish. Then just across the river there was a coal yard and, our windows fitting as badly as they did, our curtains (white candlewick if you please) needed washing every week and were washed at least every fortnight. The floors were very uneven because of the subsidence caused, we were told, by the nineteenth-century coal workings far below. The bedroom door fitted the hole for which it was made when it was closed. But when it was open it swung a good three inches clear of the floor.

When we had been happily ensconced for a few days we spotted the real reason for our bargain. The close next door was shored up by huge wooden stays that were never part of the original builder's plan. Indeed, the next tenement was soon abandoned and we had a demolition site next door. We worried remarkably little about whether our close would be next.

Anyway, the builders were soon patching up the wall and we set about making a very fine home out of it. The sitting room, which was a rectangular twenty-five feet by sixteen, we gave Wedgwood blue walls, all eleven feet of them, and a yellow ceiling, with the wonderful mouldings and central rose picked out in white. When that was done we had, with our light blue patterned Indian rug, the most beautiful sitting-room we have had to date in fifty years of married life in eleven houses. Our first decoration was in the bedroom. There we painted the ceiling a rather dark shade of greeny-blue. I remember telling Fiona sagely that a strong colour on the ceiling would lower it and make it cosier. When we showed our handiwork to my parents' friends the Dewars, who had once lived on the top floor in the same close, it was a rare case of crossed lines. Far from the expected praise, what the young couple got was, 'Oh well, dears. Don't worry,' from Mary, 'You can always change it.' That could have been a catty comment but I don't think it was. I think she really thought we were showing her the challenge we had bought rather than showing off our first improvement.

We enjoyed living up a close and are still quite proud of having done so. You can't really say you have experienced Glasgow until you have lived up a close. Once you got used to the noise it was really couthy. It was also interesting because you could hear so much of life as it washed from floor to floor. I don't remember much in detail except the poor English woman in the Dewars' old flat. She had four children and was struggling to cope. I don't know how often I heard her screaming at those little bairns, 'I shall hit you.' I am fairly sure that she never did and that that was one of the problems. And our very nice neighbours across the close, whose daughter babysat for us at least once, had most exciting rows which could be best heard through the communal wall between the two kitchens. The argument would rise to a crescendo, whereupon the daughter would crash along the corridor and bang out of the door to be followed by the father in an earnest endeavour to haul her back by the hair. It was very exciting but foreign to me, brought up in Aberdeenshire among people who were more inhibited and more thinly spread.

Another feature of urban life to which I was introduced in Glasgow was the Indian curry house. There was one on Bank Street, very near the Students' Union and I got my first dozen hot lamb curries there. They were unlike any food I had had before and I got on well enough with the staff to get into a sort of a competition with them in which they tried to make me a curry so hot that, though I might manage to eat it, I would suffer a fiery mouth. When he delivered his hottest curry vindaloo the waiter said, 'Very good for married man – not so good for single man.' I never knew what he meant exactly, though I knew it was a piece of manly chauvinism, a joke and probably obscene.

We liked Glasgow but we were soon looking back on it with affection. Within the year, my research finished and well on the way to being published in the top journal in the trade, I was ready to try the market. I thought I would get on better in Economics rather than Economic History so I tried for a step up in Dundee, at Queen's College of St Andrews University. That brought me into contact with a very strong character called Archie Campbell, a man who may have had friends, but who certainly had enemies. In his dealings with young Allan he won the first round. I asked for a lectureship but was offered an assistant's job with no increase in pay or status, and I took it.

So we sold the flat we had bought for £700 for no less than £1,150 pounds, a profit in a year of 50 per cent if you don't count the fact that we had painted the whole place outside and in. We moved at first to a university flat on the Dundee campus and enjoyed a very different and much more working-class standard of urban life. We were just off the Hawkhill, notorious for jute lassies and riotous living. There were some very good things about living there. The wee wifie in the tenement off the Hawkhill who kept the tiniest shop in the world – open all hours –was one of them. Her shop had a terrific turnover per square foot, but the trouble was there were hardly any feet. The tiny shopkeeper was completely hemmed in by her stock, which went from floor to ceiling. She knew everyone from the Hawkie and everyone from the flats and among the students. When the young lecturer went in to get a quarter of a pound of tea and a packet of digestives in his first week in his second job after graduating she welcomed him with a warm, 'How are you getting oan, son?'

'Oh, fine,'

'It'll be a big change from school?'

And I joined a catholic clientèle in the Marine Bar, the oddly named pub (for it was a mile from the sea and perhaps 300 feet above sea level). There Bob the host kept order among his clients with the aid of a knobkerry which one of them had brought him from Africa. I am sad I wasn't there on the following occasion, but my colleague Peter Henderson was. This Rangers supporter came in looking for drink after the Huns had beaten Dundee at Dens Park, just up the road. As the Blue Nose shoved through the double saloon door Bob went for the knobkerry and hit him so hard on the side that Peter swore he had broken his arm. As the visitor, who had now guessed he wasn't welcome, was trying to get his undamaged hand out of his coat pocket to open the door and withdraw, Bob, who claimed he thought he was going to produce a gun, broke his other arm with another mighty swing of the knobkerry. You can't get experience like that in Methlick.

Soon after I arrived in Dundee I was done the great honour of filling in for Professor Campbell with his nine o'clock lecture. It is a nice part of the Scottish tradition that the top man usually takes the first-year class so that the tender plants don't have the young shoots of learning manured with the wrong stuff. Sadly, the class were like the old wifie with the wee shop, they didn't believe I had come to lecture to them. The noise level remained high. Conversations were animated. No one looked at me. If they did they soon looked away. The young lecturer cleared his throat. Nothing was heard. He was desperate. What on earth to do? He couldn't just leave or that would have been it – the shortest lecture career ever.

Well, not quite the shortest, for just before I got to Glasgow University one of my fellow assistant lecturers there had a horrendous experience in his first lecture. His brief was to inform about 200 engineers about Economics. Not a bad idea, you may think. Engineers who have no concept of what things cost can do a lot of damage. But these young lads didn't want to do Economics and they did not have to pass an exam in the subject. They had to attend, but they did not have to pay any attention and most didn't. They preferred to act the goat.

Anyway, my friend was extremely nervous about his first lecture but eventually screwed up the courage and rushed in, read through his notes as quickly as he could, slammed his folder shut in such a way that his notes fluttered about and several pages landed on the floor. He eventually got them scooped up and dashed out – into a cupboard. He decided to stay there so that he would avoid adding to his already considerable indignities. After several minutes the hubbub in the lecture theatre had died down. He tentatively opened the door of his cupboard and looked out, to be met by 200 paper darts. The students had known the score exactly. And had made their paper darts as they waited for him to re-emerge.

The lecturer didn't leave immediately but took the first good offer of a job in the Treasury and disappeared. I did not want my career cut so short. Besides, I was not nearly as bright and there would be no juicy job in the Treasury waiting for me if I couldn't handle 200 first-year students in Dundee.

I eventually grabbed the ten-foot pointer which was for use on the great overhead blackboard, and brought it crashing down on the desk that ran across the room between the students and the lecturer. It was a really loud crack and they were shocked into silence. I grabbed the chance, apologised for Professor Campbell's absence and launched into my lecture on 'the gains from trade'.

I did quite a few of the professor's lectures for him and everyone knew that that was quite an endorsement, so we really got on well, though every now and then Campbell let himself down very badly. Just after John F. Kennedy was assassinated he introduced a visiting American professor in the following terms: 'Professor So and So is a specialist in money but thank goodness he isn't going to speak about that today . . . so as somebody said to Mrs Abraham Lincoln at the theatre after her husband's assassination, "And apart from that Mrs Lincoln, how did you enjoy the show?"' Knowing him, I really expected Campbell to go on to say, 'So Jackie, apart from that, how did you enjoy the drive through Dallas?' but what he did was nearly as bad. He said, 'Of course that story has been brought up to date,' and then he giggled. His assembled staff were horrified and I am still horrified.

But this boor did me a favour. When he fee-ed me he said that I and the other four assistants would be promoted 'as and when' we had two articles published in the journals listed in the *Journal of Economic Abstracts*. Now that may not sound much, but he himself had not managed a single such act of scholarship and he had made it to a chair. So, though he was asking a lot, I quite liked that. It was a challenge and it was clear. I have always worked best under pressure. Also, I had already finished my work on the Glasgow Improvers which my assessor had said was 'brilliant', so I wasn't daunted, even though not one of the staff of ten or so at Dundee at the time had got anything in that journal. By the end of another year I had done a piece of research into something I had called Fiscal Marksmanship and it had been accepted for publication in the *Oxford Economic Papers*. Professor Campbell confirmed that next year I would go up to lecturer.

Now comes the bitter bit, and it doesn't reflect well on the then heir to the little farm on the hill. At the end of the year my promotion was announced, but so was the promotion of the other three assistants, none of whom had published anything – not in the *Journal of Economic Abstracts*, not anywhere. I should have risen above it. I tried to tell myself that there was a shortage of qualified economists and they were all able enough. Anyway, I went to see the professor and reminded him that I had had a year's seniority over the others and I wanted to retain it. During this year, when I was the only one who had done what I was told, I had gone from being senior to the others to being on a par. I told him I thought it was fine to promote us all but what had I done to lose my year's seniority and fifty quid? As the drift of my argument unfolded the great man grew pink and then dark red in the face. In an absolute fury he stammered out through clenched teeth, 'You you you come here and and and demand and . . .' I forget whether I heard the rest, for I had a kind of queer itchiness behind my knees that told me I was about to get to my feet, ready for action. Even as I rose (watching myself as it were from ten feet up) I wasn't sure whether I was getting up to plant him one across the great desk or whether I was rising to leave. Luckily it was the latter.

CHAPTER SEVEN
To the Hills
1964–65

I first met Bertie Paton in 1964 when I was lecturing in Dundee. With the family, I had travelled the twenty miles to Cortachy Highland Games in the grounds of Lord Airlie's castle, which boasted its own cricket field on which the Games were held. Bertie was making a below-average job of the throwing events but was clearly a strong man, and, while not really tall enough to be an ideal heavyweight, was well on his way to being more than heavy enough. He told me he lived in Linlathen, not two miles from where we were at Dalclaverhouse in Dundee, so we arranged to train together while I would do my best to pass on enough knowledge to make him not quite good enough to beat me. It was the start of an extraordinary friendship. I taught him what I could about Highland Games, and I encouraged him to see beyond being the best ganger with Betts the Builders to a house of his own, stocks and shares and his own business. He had already taken giant steps in the right direction by successful courtship of the only daughter and remaining heir to the most beautiful hill farm among many in Angus, Spott in Glenprosen. He taught me a lot of what he knew about the art of life, and not all of that was the sort of art of which my granny would have approved.

Right away Bertie Paton got me into hot water. To assist with the caber-tossing classes, he bought a redundant telephone pole from the Post Office, as BT was called in those days. So after training we went up Glenprosen to collect the pole. A couple of pints later we were loading this onto the roof rack of Fiona's little VW beetle when a very important and angry man screeched his shooting brake to a halt and accused us of being thieves. I now realise that the reason why Bertie didn't use his skill as an amateur boxer to deck him was that the angry man was Lord Airlie's factor David Laird, with whom we both subsequently became quite friendly. When Bertie arrived at what was then the home of his intended, the police were waiting. He was able to show them a receipt from the Post Office and that would have been the end of it had I not written, on St Andrews University notepaper, a letter showing that I too could be important. I told Lord Airlie that he would need to keep control of his factor and not let him embarrass folk in front of people they hoped would become their in-laws. His lordship handled me very well, pointing out, very reasonably, that it was in the gloaming just after closing time that a lot of the devil's work was done in Glenprosen and that his man should be forgiven for being suspicious.

The caber-tossing went all right and we two pals practised away on a bit of waste ground at Dalclaverhouse. Bertie quickly learned just about all that I could teach him. He won the local championship at Glenisla Games but couldn't really get far in the open competitions on account of the amount of metal he had in his legs as a result of a motorbike accident. We shared one memorable day though. We went all the way over to Arisaig on the West Coast and came home with a pocketful of tin. Those Games are held on a Wednesday, which is unusual, for West Coast Games were usually held on Thursdays, and that was the secret of our success. We were surprised to find that, apart from ourselves, the only competitors were local lads. That was strange, as Bill Anderson, the undisputed champion, had said he would be there and I had expected him to bring along other champions, including Bob Aitken and perhaps Sandy Gray.

At any rate, I duly won everything including the silver cup, and Bertie got prizes in all the heavies. In jolly mood we set off for Dundee. When we got to Glenfinnan, only half an hour along the road, who should we see, checking in to a B&B, but Anderson and Aitken intent on a good night's rest before the Games. We didn't stop to commiserate. I held my hand on the horn and, with Bertie holding the cup out the window, we sped gaily on.

Not long after we first met, little tributes of game started appearing mysteriously at the door of our little terraced house in the grounds of what had once been the Claverhouse jute mill. There was a pheasant and there were several trout. We were to find out where they came from when Bertie introduced us to life up Glenprosen. That was a life-expanding experience.

The thing that had the biggest impact in the shortest time was getting to know the Royal Jubilee Arms and its proprietors, John and Hazel Duguid. They ran an establishment that was only just credible. You would be driving up this beautiful road through mature trees on the Airlie Estate when at Dykehead, just before the road splits between the glens Prosen and Clova, suddenly there she is – all ableeze, carpark full and buzzing by day and by night. In 1965, remember, pubs closed at 9.30 p.m. But this Victorian hotel, built by the lairds to accommodate their shooters, appeared not to be subject to the laws of Scotland. Having discovered it, we could get a babysitter in and leave the three children bedded at half past seven, drive up to Kirriemuir and on to Dykehead, have a couple of pints and a beautiful T-bone steak and enjoy dinner dancing, with floor shows put on by such dazzling stars as the Tartan Lads, the Corries, Val Doonican, the Alexander Brothers and others so unforgettable that I can't remember their names. If there was a closing time we never saw it. In Dundee it was difficult even to get a meal by the time we got the children settled.

The first fish which we found on the doorstep had been caught after a Saturday night at Dykehead, when Bertie and a well-oiled team of revellers went up to the Backwater Dam in Glenisla. The reservoir was being built at the time and the boys knew how everything worked. They shut off the water so that the dam started to fill and the pools below ran dry enough to let them stock their trout bags.

Bertie showed his lecturer friend how to take bigger fish. I could tell you the location of Bertie's favourite pool, but then I would have to kill you or else he would kill me. It's on a bend in the Prosen where the swirling river has cut a deep track in the rock and made a grand lie for salmon.

He would tie eight feet of string to an ordinary snare for rabbits. The snare would then be stuck in the split end of a rowan stick about six feet long. That would then be lowered into the water. When the snare was round the fish's tail the stick would be let go, the string pulled and the fish hauled out of the water.

I've only seen it done the once but I was impressed with a fine salmon of about six pounds. But how to get it back to the car? We had about 150 yards of open glen to cross in full view of the road and the bailiffs were always about, it was said, though I never saw them. The plan was very cute. It was also the easiest way to carry a salmon. A string was attached to its gills and it was dragged along. From the road Bertie might as well have been walking his dog – which couldn't quite be seen for the long grass and the ferns.

My saintly granny could not have disapproved much of that, for when she was Mary Yull she helped her father to escape with a salmon caught in exactly that method in the River Ythan. She was on Donald, her pony, and under the noses of the bailiff, carried the fish home concealed in her skirts.

I am not so sure that my granny would have approved of our other salmon-fishing exploits. If I don't remember all the details it is because I was tired after a very jolly evening in the Royal Jubilee Arms. The captain of the team wasn't Bertie, but Stewart Macintosh, the very successful breeder of black-faced sheep from Cormuir, the next place up the Glen from the bridge at Spott. Sadly now departed, Stewart opened my eyes to the difference between a tup on the hill and a tup at Perth or Newton Stewart. I had no idea that those beautiful curly horns were the result of careful work with a file or that they heated the horns to make them soft so they could be bent into those crazy twists. Nor had I ever seen a gaff used to land salmon that had not first been tortured on a hook and line. Indeed, I thought that sort of thing finished before the Second World War. At any rate, when all decent people were in bed Stewart led his team into the upper reaches of the Prosen. It was late autumn and there was only a little ice at the sides of the burn and where the water came tumbling over the rocks. It was cold, but I didn't at first heed the advice to piss myself to provide a bit of warm insulation in my boots. Stewart and Bertie were the gaffers and Kate, who had been set to be the next mistress of Spott, the finest farm in Glenprosen, until her fiancé Archie Whyte died young, carried the bags. Well, it was all too easy. Stewartie shone his torch on the water where he could easily see the salmon and soon had one gaffed and handed to me for putting in my bag. Soon there was another and another. It was exciting, but after about five fish I noticed a funny thing. My bag wasn't getting any heavier. Horrors. There was no bottom to the bag. Stewartie shone his torch down the fast-flowing burn and there every thirty yards or so shone a white belly, making its way down to the Montrose basin. From then on both gaffers used Kate's bag and soon we had plenty fish. But not before a strange light had appeared on the water. Someone said 'doon' in a sort of stage whisper and I, forgetting that I was in the Prosen, went down to the neck into the freezing stream.

It may not sound like it, but it was good fun and the salmon were good to eat, though some were a long time from the sea and past their best. But on that trip I was introduced to something else which astonished me, and showed me that there were a lot of differences between rural Buchan and the Angus Glens. The next morning Bertie and I were out for a run up the Glen. It was beautiful day, the grouse were filling the air with their calling and whirring and life seemed very good. But I was alarmed when Bertie stopped at what I thought was a shepherd's lonely cottage and said we'd better go in and see Charlie. I knew that Charlie Milne was the head gamekeeper and I knew we still had the salmon in the boot. I also knew that if people turned nasty you could get a four-figure fine and lose all your equipment, including your car. However, Charlie was pleased to see us. He brought out his bottle and Bert went to the car for his and we had a jolly half-hour. Then as we were getting set for the off Charlie said, 'Aye and did ye get ony fish?'

'Oh aye,' said Bertie, 'Ye want a look?' and to my horror opened the boot.

The head gamekeeper took a look and said. 'I'll jist tak this een.' And lifted out a nice three-pound grilse, maybe the cleanest fish we had.

Bert tried to explain what was going on and why we weren't on our way to jail, but all I understood was that although Charlie Milne was indeed head keeper, his responsibility was grouse and pheasants. They were natives of the Glen who had to be nurtured. Salmon were no better than vagrants who passed through his lordship's property.

Then there were the hare hunts. The mountain hares, those great animals which are just like real hares any other time, but turn white when the snows come to the glens, are considerable pests. They say that four hares can eat the grass the farmers would rather give to one sheep. So there were annual hare hunts to try to control their numbers and I was invited to join the hunters at the Spott. I had been the daft laddie at the fishing, but when it came to the hares I expected to do better. Sadly, I found that I had a thing or two to learn there too, despite the money I had got from the butcher for Little Ardo hares after the war.

Our first drive was up a long slope which gradually got steeper and steeper to the top of this hill, where standing guns would polish off as many hares as possible as they fled from our guns. I got off to a flying start, or at least I thought I did. We hadn't gone a hundred yards when a hare got up in front of Bert, who was next to me. Bert was too slow to fire, let it go, and when it landed in my quadrant I shot it. A clean kill. Whar's yer daft laddie noo? Another few steps and one got up in front of Charlie Milne. He seemed slow and again I shot it. These were nice big hares full after a summer getting fat on Spott's grass, so they weighed about six pounds apiece. Soon I had shot five and my bag was nearly full while the crackshots next to me hadn't shot a thing yet. I was proud but my bag was getting heavy, though not half as heavy as it was by the time I had carried those hares to the top of that damned hill. Bertie and Charlie Milne filled their bags when we were nearing the top of the hill and arrived fresh while the daft laddie was pouring sweat. 'Aye,' said Charlie Milne, 'ye got it wrang, ye should let them cairry themselves up and shoot them at the top o' the hill.'

When I was in Dundee and for many years thereafter I used to enjoy the change from university life to doing almost anything with Bertie. Like the time I jumped at the chance to keep him company when he took his digger from nine miles up Glen Prosen to dig a rubbish hole for somebody nine miles up the neighbouring Glen Clova. We got our hole dug but I knew Bertie had a very important reason to be back to base at five o'clock. Though it was none of my business I did ask him at four o'clock how he was going to manage eighteen miles an hour on single-track roads in this heap. 'We'll gang ower the hill and doon Glen Terry,' he said, nodding at the steep hill that rises some 600 feet on the north side of Clova. I was appalled. There was no road. It looked like it was a one in one gradient, though Bertie drove his digger straight at it. I soon decided I would walk so that when it capsized I would be able to go for help. When we reached the top of the hill I was sure my pal was beat this time. There was a gate in the deer fence that ran along the top. It was about six inches too narrow to take the digger bucket and the fence was too high to lift the bucket over. What happened was similar, it seemed to me at the time, to the feeding of the 5,000. Bertie drove the digger at what passed for speed for that old machine at an angle of forty-five degrees. When the front bucket hit the first strainer, shoving it sharply over, he slammed on his side brakes and the machine leapt to the side and slewed through the gap as the two strainers snapped to attention behind him. I was impressed, because at that time I hadn't even heard of side brakes. And I was to learn more before we won down into Glenprosen. As we sank in the bog Bertie showed me once again that the impossible is only for 'them that canna dae nothing'. Without turning a hair he sent a long arm out from the digger and pulled us down to harder ground.

Yes, Bertie Paton could do things. He was a daunting friend really. He did me the honour of telling me I would have to be his best man, though he did spoil it by admitting that it was because neither of his brothers had a kilt.

CHAPTER EIGHT

Strathclyde University

1965–73

Fiona by this time had our three children as well as a house and me to look after, and I had nearly battered the Dean of the Faculty of Social Sciences at the St Andrew's College of Queen's in Dundee. We needed a holiday and we saw just the thing. For not too much money, these two sweet old ladies would rent us their little cottage on the shores of Lochcarron for a fortnight while they had a holiday. It would be a wonderful break. We would just let the children out to play on the shore while we would sip our coffees and wave to them from the window. At nights we would be snug by in our little cot and go to sleep to the gentle lapping of the waves on the shore. I bought a whole ham which Fiona boiled to perfection with loads of bay leaves and cloves.

The whole thing was a nightmare. The cottage opened right onto a busy main road. If we took the children to the beach we had to stay with them or they'd have run across the road to get us as soon as one hurt himself or one of the others. It didn't arise as the weather was too foul. With all three in nappies at night and these being the old days when mothers washed nappies, we needed some way to wash them. Washing machine? Don't make me laugh. It was one deep sink and the only source of hot water was a back boiler on the kitchen fire. But the two sweet old ladies hadn't left us any coal or sticks; indeed they must have used every bit up before they left in case we got it. The kind lady at the shop told us that, yes, she could order coal for us but it had to come from

Inverness and it would be three weeks. That was not much use for a fortnight's holiday. The nice lady did sell us one bucketful so we could get some clean nappies, and those were hung up on the wires across the road. That was all right, but when we judged they might be dry the tide had come in and our clothes were twenty yards out to sea. Lochcarron is picturesque but we were stuck. We tried gathering sticks for a bit of fire but with all the rain they were hard to burn. We were cold, we had no hot water, the beds were damp and after the second day the ham started to cry out for a refrigerator and by the third day it was turning green. No there was no one in the village who would babysit. 'The young girls just sit for their own.'

On the fifth day we went down to Stromferry where, overlooking the pier, is a fancy hotel. It was like coming back to earth. The staff were nice to us. We got reasonable terms for a long week of full board and the staff whisked the pile of nappies away and had them back in no time, smelling like a chemist's shop. We really should have left the ham on the table with an invitation to the two nice old ladies to help themselves, but we just threw it out and left the place as tidy and as clean as we could without hot water.

There wasn't a lot to do at Stromferry, but we were warm and dry and we could get a drink without a babysitter. After one night in the hotel I had to rush off to Glasgow to seal an escape route from Professor Campbell and that left Fiona and the three children without even a car. However, there was the ferry. They went back and fore on it many times a day. And on one day the heroine took her brood to Kyle of Lochalsh by train, where they got another ferry to Skye and shivered on the stony shore, looking at the Cuillins until it was time to take the ferry back to the train and return to our cosy hotel.

It was a memorable holiday and not only for its problems. I got the job in Strathclyde. Professor Kenneth W. J. Alexander, whose teaching I had so admired in Aberdeen, wanted me to work for him and he not only gave me the year's seniority that Campbell had stolen but he gave me three increments more. I heard later that there had been some discussion about my reference from Dundee.

It had said along with all the usual good but unexciting things you get in such references, the interesting phrase, 'Allan has gaps in his knowledge'. Well, of course he did, but in academic references the normal practice is to say all the good you can about the candidate and leave the bad things to the imagination. Campbell's phrase was code for 'this guy knows nothing'. Anyway, I heard later that one of the panel had asked if this Campbell was the one who had been a lecturer at Glasgow University, and on having that affirmed said, 'Well if Allan didn't get on with him that is good enough for me.'

And we weren't to know then what bonus it would be, but Fiona found when we got home that she was pregnant with our fourth child. I was twenty-six and she was twenty-eight. She told me with a finality with which there could be no argument, that that would be our family complete. She told the doctors too. She refused to leave the hospital until they had tied her tubes.

So it was back to Glasgow, this time to the University of Strathclyde. Having enjoyed ourselves in a tenement on the north side of the river we crossed the city's cultural divide and bought a whole house on the south side. We sold up our tiny new terraced house in Dundee, and in Mansewood Pollokshaws we were able to buy for the same money a huge stone-built Victorian mansion. We had a garden big enough for football though not unmanageably large, but the house was rundown. We were in luck though. The Dundee Mafia came riding to our aid as we set about remodelling and redecorating our new abode top to bottom. Peter Murchie, our architect friend, drew a wonderful set of plans with modern kitchens with picture windows bashed out here and French windows bashed out there. Peter used his contacts in the trade to get us the half-acre of cheap jute carpets and curtains for the giant windows, and at weekends Bertie Paton brought his gang down to do the structural work. Better than that, they brought remarkably cheap materials. The windows were from Bett's the Builders, damaged stock. (What are you needing damaged this week Bert?) That summer the squad came down on the Friday, tried to drink the Coach and Horses dry of Guinness, and on

Saturday they worked while I went off to the Games to try to win enough to pay them. And Bertie was able to further my education in how things really work when we ran out of cement. He sent me up to the new school that was being built just round two corners to get what he needed. And sure enough the night watchman ran a shop all weekend. 'Form an orderly queue and have your money ready. Sorry, no receipts' No wonder that school, and all the others, ran over budget.

I can't go over it all but I got another memorable lesson from Bertie on how some people can do things, and that I'm not really one of them. There was in what had been the kitchen, an old-fashioned steel range, one of those black monsters which generations of kitchen maids had to polish every day with black lead. We wanted to remove it and decided to break it up rather than extracting it in one piece. Well, as the champion caber-tosser I volunteered for the demolition job and set to work swinging a twenty-eight-pound throwing weight. It was stronger than I had thought so I redoubled my efforts until a small piece fell off, and then swung again and another small piece tinkled to the floor. Bertie watched his friend sadly until he could stand it no longer. 'Get oot o' ma road,' he said without due regard to the champion's feelings. Then he gave the great steel monster four or five sharp cracks with a four-pound hammer and it simply fell asunder.

It was in Glasgow on the South Side that the farmer from Aberdeenshire first became aware of religious hatred. This was a great surprise, for which life in the North-east had not prepared Fiona and me. I had only come across those who believed and those who didn't. But in Glasgow people were hating one another in the names of their religions. I was never aware that the Jews were involved in this. And there was some slight disregard for brown people despite the great service they gave with their little shops, but that I blamed, not on religion, but on race. It was the Christians who were the problem. Catholics and Protestants hated one another. We had always known about Rangers and Celtic at the football but I had had no idea how serious it was. In that seat of learning, the University of Strathclyde, the plumbers who came to put in our new kitchen as a homer were all Catholics and the electricians who rewired the house were all 'Prodisents'. Why? I was told, 'Aw, naw. It just wouldna work to mix the faiths.' When the first big Orange Walk came down Thornliebank Road with its flute and pipe bands and perhaps 10,000-strong I rushed down with the kids to wave and clap them through. Wasn't it great just to be marching along maintaining your traditions? It only dawned slowly that they were also marching to nurse their hatred for fellow Christians.

Just how naïve we were about bigotry can be seen from the way we handled the problem of choosing a school. We really didn't want our five-year-old treasure to be beaten and all the state schools around us were ardent beaters. However, we did find this very nice convent school where the Mother Superior was clearly a gentle soul who would not have gratuitous violence and she didn't mind that we weren't Catholics. We had no idea that we were doing anything abnormal but we were to find out. That Mother Superior had left by the time Sarah started at the convent, which was a pity. Every Monday morning all those who hadn't been to mass had to stand up to be humiliated by the teacher. When we heard about this we should have removed her immediately, but no: I found the nearest chapel and took her to mass the next Sunday. The next Monday Sarah went off excitedly to school. When the teacher told all those who hadn't been to mass to stand up Sarah sat proudly in her seat. 'Sarah Allan, why aren't you standing?' 'Because I have been to mass.' 'No you haven't. Go and stand in the corner.' Life is hard enough without that sort of thing. We found a kind school and sent Sarah there. That was a success but one of her friends, home to tea, told us that she hated Catholics. 'And have you met a Catholic?' 'No!' she exclaimed, and she shivered as though she had found something slimy beneath her pillow.

Strathclyde University was an exciting place to be in those days. It had just been made up to a university from the old Andersonian Institute of Science and Technology and there was a lot of enthusiasm and optimism about. No job seemed to be too much, so many were offered and I was able to take some up. I got work from various government departments reporting to them on what was happening in the fields of industry and public administration. By far the most interesting of those was when I was sent down among the Clyde shipyards to report on life at John Brown's. Well, if I was out of my depth with Bertie and his pals up Glenprosen, I was twice out of my depth among the tough guys of the shipyards. I had to sign the Official Secrets Act, so I am not supposed to tell you anything about it, but luckily I can't remember much, except that the whole business of running a shipyard was jaw-dropping. I had to interview all the shop stewards individually and in groups and report on whether the productivity increases they were citing to earn pay increases under the Prices and Incomes Policy were real or not. Well, what in hell did a laddie from the farm, whose only training had been behind a desk and splashing around in the Prosen after salmon, know about building a ship?

Anyway, what I did learn was just how little running a shipyard with several thousand men resembled running Little Ardo. And I did admire the shop stewards. They put me on the spot right away. I was introduced to the top stewards at a meeting. There were about twenty of them, all tough guys, and one started off by asking me to explain who I was and what qualified me for the job of commenting on their pay. I was taken quite by surprise but managed to handle it somehow. And I got on well with the stewards, though the biggest problem was to get them to stop telling me stories and start telling me the story. I can only recall one and it was told me by a boilermakers' shop steward by the name of William ('Silver Tonsils') Connolly. It's a Jeelie Piece story about a wee boy shouting up to the tenement flat window, not for a jeelie piece, but to report 'Mammie, I couldna get it. The chemist was closed.' 'Awa ye daft wee bauchle,' the mammie replies, 'I says "a fresh lettuce".'

I also had to interview all the foremen whose job it was to hold the ring between all those tough guys who knew how to weld, fire shots and apply red lead, and the managers who didn't, and often didn't know any more than than the twenty-six-year-old laddie asking questions for the Prices and Incomes Board. One of the foremen's difficult jobs was keeping the boys at work until stopping time. Your place of work in a yard could be up to half a mile from the gate, so there was every incentive to make your way there early. Five minutes later at the gate could mean an extra twenty minutes before you got a pint, which accounted for the iconic rush of humanity out of the shipyard gates at five o'clock. Anyway, I was told about some new college boy in management who hit on the bright idea that, half an hour before knocking-off time, the foremen would make their ways to the gangways of the boat and wait there to stop anyone nicking off early. This was a predictable disaster. The boys didn't like it and retaliated by, as soon as the foremen left the job, sitting down to read the paper or have a hand at nap. It was like Bertie Paton's story about the building site in Dundee where, despite Herries fencing and padlocks all round it, they were losing an awful lot of materials. So the management set up a checkpoint at the gate and searched everyone. Along comes this bright spark with a wheelbarrow half-filled with loose straw. AHA! This was clearly what they were looking for. They stopped the man and searched the straw but found nothing, no bathroom fitments, not even a bag of cement. Several times this suspicious character made his appearance, was searched and allowed to proceed before they twigged that he was stealing wheelbarrows.

Sorting out a policy on wages consistent with the government's aim that wage rises should only be given where increases in productivity could be shown to justify them, was made impossible by the truly bizarre maze of wage rates. There were over 300 rates in the yard. Each trade (and each grade within it) had a rate. You got an extra one tenth of a penny per hour if your father had worked in the yard. And there was extra for working when it was too cold or too hot or too windy. There was danger money, height money and dirty money. I had it from both sides, so I think it was true that the biggest rate was shit money. You were not expected to work among human excrement, and with such big distances to the toilets people did get caught and have to squat down in the double bottoms of the ship's hull. So if you found a turd you had to report it to the foreman, who would tell you to get the shovel and bring it up, whereupon you were credited with £1, almost three hours' work for a boilermaker at that time. The boys clearly enjoyed telling me about that, so I am not so sure of this ground. But I was told that some did deliberately lay turds in secret places to earn shit money but also about something called 'splittin the shit'. If you got a big enough turd you could just bring out half of it and tell the foreman you'd discovered another, in which case he would tell you to bring it up and earn another £1.

That was interesting work, as was the six months I spent reviewing for the Herring Industry Board. Those jobs were paid extra and they fairly expanded the horizons of the heir to Little Ardo farm on the edge of Buchan. I got the jobs through my teacher and now boss Ken Alexander, who was himself trying to sort out wage rates at the Upper Clyde Shipbuilders with Jimmy Reid, Jim Airlie, those even tougher tough guys.

I got plenty of interesting teaching to do and Alexander even got Penguin to commission me to do a book, based on my lectures on taxation to the honours classes. As in Dundee, I also got to fill in for the professor when he was away, which was often, and I grew to like big classes. They are easier really because you cannot spend much time on those who don't understand or don't agree. You just have to treat them as a crowd to be entertained for fifty minutes. There was one bad moment. I had been given the engineers' class because it was a nightmare and Alexander thought I was less likely to be devoured by those noisy youths. It was a ridiculous class really, because, as at Glasgow University, attendance was compulsory but there was no exam to be passed. The lectures had to be endured but there was no incentive beyond basic politeness to pay any attention. I came to a sort of accommodation with these students. If they would behave for fifty minutes I would tell them a joke and then they could leave at five minutes to the hour. That worked quite well. I got a reasonable hearing, but as the joke might take five minutes they started to get restive at ten to. Then they started to get restive at a quarter to the hour. They would shuffle their feet and then start stamping, so I let them have a particulary long story slowly told and let them out at an almost respectable seven minutes to. But they smelt blood. Just after halfway through the next class they started stamping their feet in unison. I paid as little attention as I could but soon the noise level rose to a cresendo. They all stamped in unison. We were in the old Andersonian building and it was shaking. Three floors below a meeting was taking place to confirm Allan's elevation to a senior lectureship at the early age of twenty-nine and I had visions of the plaster falling off the wall. Now I had a stroke of genius – or at least a very good idea. There was a sort of chant on the hit parade at the time, and I started chanting that in time to the stamping, 'They're coming to take me away hehe haha hoho . . .' They all laughed, gave me round of applause and settled down. I was relieved.

I was a bit disappointed by one aspect of the job at Strathclyde. I had really looked forward to seeing a lot of Professor Alexander, as I had done as a student. He was a very bright man, very well-read indeed and full of ideas and enthusiasm, but he hadn't written very much. I had looked forward to long conversations and getting his take on the great problems of the world and the small problems of my research. But he was just too busy, as indeed, with all the work he was putting my way, was I.

CHAPTER NINE

Highland Games
1959–72

To get back to my sporting endeavours, Charlie Allan won four medals in the university sports in 1960 – the hammer, the shot putt, the high jump and the hop, step and jump, but that was under my own name. There were further embarrassments with Ivor Smith, my Saturday name at the Highland Games. He improved steadily and started winning prizes in the hammer-throwing and tossing the caber as well as the sprinting and jumping. Ivor even had his photograph in the *Press and Journal*, winning a heat of the 100 yards in front of the royal party at Braemar. It wasn't Charlie Allan being banned from the Olympics, or even the University Sports that troubled me now. Ivor Smith might be banned from the Highland Games, which had a rule that no one who first entered the Games after 1947 was allowed to use a pseudonym. So if I was eligible for being thrown out of King's College playing fields for being Ivor Smith, I might also have been thrown out of the Braemar Games for being Charlie Allan. I had my excuse ready, mind. If I was challenged I was going to tell them truthfully that I had first run at our local Gight Games held in the Castle Park in 1947.

There was a good reason for banning Noms-de-Games. It had nothing whatever to do with the Aberdeenshire Highland Gatherings in which I grew up. It was to do with the Border and Fife Games, pedestrianism, whose blue riband is the Powderhall Sprint, where a man can win a year's wages for coming first, and there are bookies. In these southern Games the main thing was handicap running and the betting thereon. There were a lot of fly men around and the bookies needed to be sure that the runners always appeared with the same name so that they could keep proper records and assess form.

But my granny was right again that only the truth tells twice. I killed Ivor Smith off in 1961. No one said anything. I wasn't good enough for the Olympics anyway and I didn't go in for races with bookies, but I was sad to see Ivor go. And Jimmy Miller, the garrulous commentator, was astonished at Inveraray Games when the lad he recognised as Ivor Smith appeared as Charlie Allan. Inveraray is a midweek Games and he thought perhaps I was using a false name because I had taken a sicky and didn't want my boss to read about Ivor Smith in the paper.

So, I was now officially a professional athlete and remained so until I quit the circuit some fifteen years later, when the demands of working with my hands on Little Ardo left nothing in the arthritic bones for serious sport. But though I did win over £1,000 in 1972 and toured the world as a paid athlete, there was only one year in which the label 'professional' was at all appropriate.

My parents, who had supported me to the tune of £8 a week when I was at university and given me a roof over my head until I moved in with Fiona, considered that in 1962, when I finished my last exams, they had done their bit. My £8 a week stopped without even a finishing bonus. I was surprised but not indignant. Fiona had long since given up her job to be a mother so we had no other income than the flat we had bought and done up for let in Aberdeen, which brought in £30 a month, and what I could earn at the Games.

Now, most of what appears in this volume is just the whole truth as I remember it, though of course I don't remember it all. But in the case of my earnings at the Highland Games that summer I can tell you it all exactly because I wrote it down in what is now a very tattered little notebook. In the summer of 1962 I competed in twenty-seven Highland Gatherings which netted me £239 7s 0d including £4 4s 0d for an appearance on television and £40 for appearing in a stage show at the Kelvin Hall in Glasgow, about which more later. In the busiest week I competed in five Games from Inveraray in the south-west to Dunbeath in the north-east corner of Caithness. My best day was at Aberdeen, where I won £17, and my worst was Auchterarder, where I only won £1. In round figures I had £20 a week off the Games, which saw us through the summer towards my first pay-packet on 28 October. That pay-packet as an assistant lecturer at the university was £52 for the month, so you see my summer earnings at the Games were not to be sneezed at.

I can't tell you all of what happened at the Games nor even all that happened to me in my twenty years as a competitor. My plan though is to tell you, as accurately as I can remember, about a couple of memorable expeditions. The first is the time that Jim Gibson, a plumber from Aberdeen City and a very able and versatile runner who could do anything from a sprint to a hill race, went to a wee Games in the north-west corner of Sutherland. The Assynt Gathering is held in the fishing village of Lochinver. It was quite a safari and surely did not fit the image of professional athletics.

We were living just outside Aberdeen on the wrong side for Lochinver. To economise we had moved from the grand dower-house with its chaises longues and brass bedsteads to a cottar house at Banchory-Devenick. Bill Anderson, the Heavyweight Champion whom I knew from the Highland Games, was dairy cattleman at Cowford Farm and I went over from Drumduan to train with him one night. He lived in a fine two-storey house by the roadside and he pointed out the small cottage next door. It was uninhabited. So I went to see the farmer, Andy Anderson, Bill's brother, and managed to do a deal whereby I would work for three hours a week at Cowford in lieu of rent and we could move in. It may have seemed a big comedown from the dower-house to our little cottage with its cement floor, its tiny rooms and its inch gap at the bottom of the door which allowed free access to the mice. But it didn't seem a comedown to me. We had two bedrooms, a bathroom, a kitchen and a sitting-room, and fields all round us, though we were only two miles from the town. In our sitting-room we had a coal fire, a plastic basket chair each and a nice Georgian round table, off which for some reason, my mother had cut four inches from the legs. After a few months we got a split-new settee that would accommodate two guests, even if one of them was Bill Anderson. We had a wireless on which we listened to 'Saturday Night Theatre' on the Home Service. There was electricity throughout, so we could move the electric fire to wherever it was needed. Two special highlights of being at Cowford were being invited through to the Andersons' to watch their telly, and the fantastic high-energy teas with which Francis Anderson entertained her husband before bedtime. I will never forget Francis's butter icing. In the summer evenings Bill was able to teach me how to throw all the Highland Games equipment in the field right beside the houses, and Francis, who was from a big family and already had one of her own, was a great help to Fiona about what to do with a baby. But the big clincher was that by giving up the tenancy of the wing at the dower-house we were able to save £3 a week. When you consider that Fiona kept the house on £2 5s 0d including the electricity, you can see that £3 was a considerable sum.

Getting to Lochinver meant going through Aberdeen anyway, so I picked up Jim Gibson, with whom I was to share the expense of running Fiona's little car, at six o'clock, and we set out with an AA map, our kit and a bottle of lemonade, and stopped at a bakery in King's Street for a dozen butteries. You see the professional athletes were very careful, even then, about their diets. Then we headed west for Inverness. In those days there were no bypasses. We had to go right through the centres of Inverurie and Huntly. The main road north went right down the High Streets of Elgin, Forres and Nairn (as it still does at Fochabers), and when we got to Inverness there was no nipping across the Kessock Bridge to the Black Isle. We had to go right through the middle in the rush-hour. Nevertheless we made good time and were through Inverness at the back of nine and skirting the Firth to Beauly. Then we dashed along the north side of the Cromarty Firth to Dingwall and over the Struy by

Scotland's loneliest pub at Aultnamain, where the landlord once told me that he had a big local trade . . . 'they come out of the hills that would scare you.' Well it must have been scary because there wasn't a house within sight of the place, just flat bog with not so much as a sheep. Then it was down to the Dornoch Firth and the transport café which always tore at the imagination with its sign proclaiming 'Scotland's Premier Roadhouse'. Not eleven o'clock yet and we were almost there. Look at it on the map. From Bonar Bridge it is only sixty miles west-north-west. We would be in nice time for the start at 'one o'clock or thereby' as the secretary had told me on the phone.

And so we might have been, had I not become too sure of myself. I noticed that we didn't need to take the main road down the side of Loch Assynt. There was a much more inviting road round the coast to the south. We decided to go that way and see the great mountain Stac Pollaidh and the joys of the metropolis of Inverkirkaig. It was not a wise move. The road to the west was bedevilled in those days by long stretches where there were passing places designed so that slow-moving traffic could pull in and let those in a hurry, like professional athletes, get past. That was a great help on roads that were already becoming too small for the traffic. But the coast road was so basic that it didn't even have formal passing places. It wasn't even tarred over long stretches and the traffic had to proceed at the speed of the slowest most nervous caravan driver, and he was slow. There was no need of a warm-up when we arrived. We were already boiling over.

Despite all the hold-ups, we arrived at the An Culag Park beside the harbour in time for the unique opening ceremony. The chieftain, not necessarily a laird but a local worthy, elected to the job each year, arrived on the chieftain's barge, a trawler gaily festooned for the day with bunting. He was met by the Pipes and Drums and marched up to the field. Then it was on with the Games and I had what I think was my busiest day on a Games field. I competed in fourteen events, taking a prize in all but three of the eight strength events. I can't be sure of exactly which those prizes were, though I did win the 100-yard and 220-yards sprints and the high jump, and I can tell you that my prizes came to the princely sum of £11.

They included a famous battle with Sandy Sutherland of Alness, even above Jay Scott, the bonniest man to grace the Games in my time, and a fine hammer-thrower and putter. He already had the first prizes in those events under his belt and was everybody's favourite to add the weight over the bar to his total, but I put up a brave fight. Indeed, it seemed that I had pulled off a remarkable first equal when we both cleared the top of the stick at 11 feet. Sadly somebody had a bright idea. He went down to the harbour for some fish boxes which were used to place the stands on. Each fish box, when placed under the uprights, raised the bar about eight inches.

The announcement of the winner did make us laugh. It was 'Charlie Allan 11 feet and one fish box. Sandy Sutherland 11 feet and two fish boxes. Sutherland the winner – well done Shandaig.' The athletes then had their traditional warm-down, a bombardment of nips in the committee tent while nice old men told me my time to win the heavies would come, as a prelude to telling me how good they were when they were my age, then in the pub with my old headmaster from my mother's school, John Wilson, who was camping by Loch Assynt with his very attractive wife Isa. Then we heroes started home on the journey which had taken seven and a half hours of driving on the way up. We'd had no lunch, so by the time we got to Scotland's Premier Roadhouse we were ready for our usual post-competition nutrition: lemonade and a fish supper with a single black pudding. After such a day and flush with £11, I decided to have two suppers. I was recognised in the shop and, knowing that the heavies were terrible men to eat, she gave us two fish per supper. So we had four fish each and a black pudding. There wasn't a crumb of a chip left. I so wish I could eat like that now, and that I had eaten more slowly then so that I could savour that feast still. Instead, the fodder was crammed in as we sped homeward and made it to Aberdeen before midnight.

It was quite a day, looking back, but we thought little of it at the time. Such big days were quite common. Like the dash from

Tobermory on Mull to Dunbeath in Caithness. The ferry got us back to the mainland after five hours of competition, at about eight o'clock. Then we had to dash up the road, find a B&B, for we never had anything booked, and try to find somewhere within reach the next day of the little Games which lies just fifteen miles from John o' Groats. After five hours of competition below the cliffs of Caithness, it was another dash down the road and another frantic search for a bed far enough south to be within reach of Lochearnhead the next day. I don't remember ever feeling as tired then as I feel now just writing about it.

One of the joys of going round the Games was meeting Scotland's landladies. They nearly all operated outside the law and beyond the reach of the taxman. There were very few signs up proclaiming B&B, because they didn't want to be taxed, but you got to know where to look. And if you guessed wrongly the wife would almost certainly know someone who took in lodgers on the fly. There was Mrs MacConach at the forestry cottages at Arochar for example. She was a very good sort, who kept a clean house and gave us enormous fried breakfasts. She was always at it, so much so that her husband used sometimes to give her ten shillings when he left for work in the morning to make sure his bed was still available when he got home. The first time we went to her Mrs MacConach quoted us a guinea for our B&B and we were worried that it was so expensive. We needn't have worried. It was a guinea for the pair of us – 10s 6d each. You couldn't charge money like that and pay tax as well.

Then there was the wonderful stay we had in Skye on our way to Portree to their Games. The athlete who on that occasion shared the expense was Sandy Wallace, the enormous policeman from Strathmiglo in Fife. Ek, as he was more widely known, dwarfed the growing lecturer in Economics at the University of Strathclyde. At the Games the next day we would be mistaken for father and son although there would not have been five years between us. The arrangement was that Sandy would motor through to Glasgow, leave his car at Strathclyde and we would drive up the night before, get a B&B somewhere on the island and so be fresh for the Games on Thursday. At the appointed hour, just after dinner time, he arrived at my office, larger than life at six foot two, admitting to nineteen stones, and eager.

'So this is whaur ye work is it, Charlie? Aye and that's aa yer books? And does that phone connect to the outside?' I told him that as a senior lecturer I was entitled to an outside phone but that I had never tried Australia yet. I was anxious to get going, not least because we had nothing booked for our night's sleep, and we had a hundred and ninety scenic miles to go and two ferries to cross before we could start looking. But Ek was interested. 'And whaur do ye dae yer lecturing?' he asked in his strong Fife dialect that went up at the end of the sentence. I explained that most of my classes were in groups of less than a dozen and they were usually held in this office.

'Oh aye, that's what all the chairs'll be for.'

'But the big lectures are in lecture theatres.'

'Oh aye. And whaur are they – could I see wan o thir?' So I showed him where I lectured to the first-year class of a couple of hundred. I wouldn't say the policeman was impressed, but he was very interested.

'And there'll be a library, or dae ye jist use the public library?' I took him to see the library and showed him the edition of the *Oxford Economic Papers* to which I had contributed an article headed, 'Fiscal Marksmanship – by Charles M. Allan'.

'And whaur dae ye get yer dinner, Chairlie?' I took him up to the staff club and showed him where we got our lunch, explaining that the helpings were not sufficient for a caber-tosser, so I usually had two three-course meals. And I explained that we had morning coffee also in the club and sometimes we met for a five o'clock swill after work. 'And what about sports, is there somewhere you can train?' I explained that the university playing fields were five miles away at Steppes, but that there was a gymnasium for weight-training. 'So could we see that?'

Finally Ek was satisfied and we set off for Skye. He settled down in the passenger seat with a satisfied, 'Well, Charlie that was grand.

And now I can say I wis through the university. There's no a lot o polis in Fife can say that.'

We sped on out through Anniesland past Bearsden and Dumbarton and right up Loch Lomondside past Jay Scott's Island and the Black Hamlet of Luss, to Crianlarich and the Road to the Isles. The Ferry at Ballachulish was busy with holidaymakers and there was a half-hour delay before we got going again through Fort William to turn left for Skye at Invergarry. It was a beautiful day, the second-last Wednesday in August, a perfect day for a drive through the Highlands. But in those days before the Skye Bridge the next part of the drive was for most an apprehensive one. Would the queue for the ferry at Kyle of Lochalsh just be awful, or would it be so bad that they would queue for ages only to have the service withdrawn at sundown? There was not much hope of a B&B if you started looking at sundown. But Ek and I weren't worried. We were seasoned travellers who were privy to a secret known to few non-locals. There was another ferry. Much smaller, much cheaper, with a much shorter crossing and almost never subject to queuing. Fourteen miles before Kyle and just past the Five Sisters peaks take the left fork at Sheil Bridge. Fork left again after a mile and that will take you over the pass to Glenelg where, in summer times only, you can still get the Glenelg Ferry. Bridge or no bridge, the wee ferry remains competitive. It was nine o'clock when the athletes landed at Kylerhea on Skye.

Then it was just a question of knocking on doors until we got someone to take us in. We knocked on every door, but everyone was already full. We didn't even bother to knock where there was a B&B sign, but even the tax dodgers were all full. We ventured further and further from the beaten track until, well south of where we had intended to be, on the River Ord, we came across a little cottage up a very minor road. The lady there was very pleased to help us. Soon we were being entertained to a dram, which we were able to return. I asked where her man was.

'Oh, he's out. Well, he's fishing. Well, poaching actually.' That was fine, but we were careful not to mention Ek Wallace's day job, as poaching is a serious business which can lead to the confiscation of your catch, your rods and even your car. When our hostess asked, he said he was a civil servant and that seemed OK. In the morning the man of the house was home and took us to the pantry to see his catch. He offered us a nice sea trout of over a pound and a half and the missus cooked it for our breakfast. I often wonder what that good woman made of us when she read in the visitor's book that Ek's home address was The Police Station House, Strathmiglo, Fife.

On the Thursday morning it was just a leisurely forty-six-mile run up the road, past the Cuillins to Portree where the Games are held in a disused quarry. It has lush turf on the floor and the crowd sits above the action on the rocky sides looking down into the sheltered anchorage of Portree Harbour. It is a wonderful place to compete, for the crowd get such a good view because they are so near to the action. The sides of the quarry seem to reflect the noise and a ripple of polite applause sounds to the athlete like thunder. The park is so tiny that they can't hold a light hammer-throwing competition. The running track is only 130 yards round. It must be the only track where competitors have to do more than three laps in a quarter-mile race, and they say milers have been known to get dizzy.

We had a delightful day at the Games. Ek won the throwing the twenty-eight-pound weight; I won the high jump. But we had a rare battle in the caber, which caused an English tourist quite a moment of anxiety. In my determination to get up as much speed as possible I overran the expected mark, and as I approached the crowd at speed this lady took too big a step backward and, according to the paper, would have landed in Portree harbour had it not been for the willing hands of her fellow spectators. It was exciting. I am surprised and delighted to say that the Health and Safety police still haven't got the Portree Games sterilised in some sports field outside the village.

Ever since I was old enough to understand these things I always wanted to win the Chieftain's Trophy at the great Aboyne Highland Games. To qualify for that you have to win at least one point in the heavyweight events and at least one in the running and leaping. The winner is the one who has the most points among the qualifiers. The senior trophy at Aboyne is the Dinnie Trophy for the heavyweights, but the Chieftain's Trophy is regarded by many as the blue riband as it truly reflects the tradition that the Games athlete should be an all-rounder. It was won for seven consecutive years by Jay Scott of Inchmurrin, the biggest island on Loch Lomond. Then in 1957 it was taken by Bob Aitken of Auchenblae, who held it for ten years until, to everyone's surprise, I won it in 1968. It was a good win too, because I joined Jay Scott as one of the two athletes to get first prizes in both sections, even if those prizes were only halves. I was first equal in the high jump and first equal in the caber.

To say I was pleased would be a gross understatement, but in Highland Games you are not expected to punch the air, hug people, run to the crowd for their adulation or to cry. So my main feeling was of panic — how would I contain myself and avoid letting myself down? I had to seem pleased but not bumptious. I had to insist that the better man had lost. Whatever was happening on the inside, I had to remain cool on the outside. That is how it should be at Highland Games and that is how I was.

However, I had come up from Glasgow in the train and when I got back to its anonymity for the return journey, I cried like a baby. Now, the Chieftain's Trophy comes in a large solid wooden box. I sat it on the table in front of me and let go. I must have been a sorry sight, for I became aware that everyone was looking at me sympathetically. A kind old lady came up and asked if it was the remains of some loved one in the box and if it would help to talk about it. I hadn't the heart to tell her that these were tears of triumph.

That success at Aboyne may have been the reason I started to put even more effort into the Highland Games, and a very important part of that was David P. Webster, the founding Principal of the Aberdeen Spartan Club, who has had a star-studded life in sports administration. He piloted trampolining from the circus to a formal

Olympic sport, and it was his early efforts with the Aberdeen Spartan Club which led eventually to the world's strongest man competitions appearing on the telly. He coached the great world champion weightlifter and folk poet Louis Martin, he looked after the British teams at seven Commonwealth Games and three Olympics, as well as being a strong man in his own right and world champion in strand pulling. Davie was a specialist in lifting but also in weight-training, and I asked him if I could aspire to increase my throwing distances by 10 per cent if I were to train with weights like a lunatic. He said he thought 10 per cent was ambitious but he was confident that if I followed the schedule he would give me, I would get at least a 5 per cent increase in my throws. I decided to go for it and he gave me a punishing schedule copied from those used by the Russian weightlifters who were dominant at that time. They were professionals paid by the State, but with a job at the University of Strathclyde I wasn't much worse off. I certainly had enough time to do all the training my body would take.

The weight-training was a success. I didn't try to estimate it in percentage terms, but my improvement put me up from one of a crowd to a position where I could earn my nickname, Charles the Third, because I was usually third behind Bill Anderson and Arthur Rowe. That success was only made possible by a ridiculous mistake.

The idea was that David gave me the exercises but it was up to me to do them as often as I could, but at least three times a week, and to increase the weight as fast and as far as I could. The objective was to lift 100 tonnes a week. Well, I quite enjoyed it and I was soon piling on the weight. At Bellahouston Sports Centre I used to take over the weightlifters' platform and, while my fellow gym monkeys were doing a bit of a grunt and then going over to the mirror to admire themselves, I was trying to cover the entire platform in sweat. On one occasion I was doing deadlifts, the original brute strength lift, when I noticed one of the others looking very quizzically at the weight I had on. He said in response to my enquiring look, 'Do you know what you're lifting there? That's the Scottish record.' Well, I had had no idea, but as a result

David Webster invited me to attempt the Scottish record at a Scotland versus Poland match a few weeks later and I was able to beat it by some 15 per cent. I still have my certificate to prove that my weight-training was intense. However, my obsessive dedication to this weight-training was hard work. I had to get to the target of 100 tonnes because I had told David Webster that if some Russian could do it I would do it too. I used to do my work in the morning and scoot off to the gym, cover the platform with sweat and then go home to bed for a rest. When the children came home from school they would see the curtains drawn and say to their mother and driver, 'Dad's home.' It was a great day when I finally made the 100 tonnes for a week.

So what was the ridiculous mistake that so helped my Highland Games career? Well, it was a very small mistake. David Webster's lifting budget, what the Russian Champions had done, was 100 tonnes a month. When I discovered the mistake I had just done my third week at 100 tonnes a week.

There was no crying in the train when, again to everyone's surprise except (this time) my own, in 1971 I won the caber-tossing championship. In thanks for his part in that success I gave David Webster a handsome shepherd's crook which had won the crook-making competition that year at Cortachy Games. When Highland Games are mentioned I am always introduced as the former world champion of the caber. It is easier to explain what that is, but the Chieftain's Trophy was my first important win and it has a much longer history, going back to when the great Donald Dinnie won nine events, including piping and dancing at the first Aboyne Games in 1867. The caber championship wasn't even invented until the 1960s. This time the trip home to Glasgow was just that, though I was very pleased.

One journalist asked me if I thought I would have won if Arthur Rowe had turned up as expected, whereupon somebody protested loudly in my defence, 'You canna expect the laddie to beat folk that don't turn up.' I often remember that one when I'm getting fed up of the commentators and summarisers on the telly trying to justify their existence by analysing sport to death.

Just a footnote on that great win at Aberdeen. Two years earlier at the caber championships at Aberdeen I had had a memorable encounter with that great politician, celebrity chef, game-show participant and general bon-vivant, Clement Freud. He had preceded me by about ten years at that extraordinary progressive school in Devon to which my parents had sent me all the way from Little Ardo. Freud, who was reporting the event for *The Times*, was amused by our meeting and made what I thought was a very Freudian comment. He said, our old school 'was a place much more likely to send reporters to world championships than competitors.' I thought it a nice self-effacing comment.

The journalist may have got his answer that winter when Quantas and Grants (the whisky people) together sponsored the next world championship in Australia. I went as defending champion, a role I much enjoyed, and I went well prepared by long hours in the gymnasium – but of course you can't practise the caber in winter's snows. The Scottish competitors were Bill Anderson, John Freebairn, who was once Partick Thistle's goalie and knows what it's like to play for the team that sits on top of the old Scottish A Division, and Gordon Forbes from Glenlivet, who later made his second fortune in the reconstruction of the Falklands. James Macbeath, the Caithness glassblower, refused to go because the championship was to be held on the Sabbath, despite efforts to persuade him that it would be all right because when we were competing it would only be Saturday in Scotland. John Freebairn from Kilsyth went instead. Arthur Rowe, the Barnsley blacksmith and past champion, was there this time; and then there were the Australians. The best of those were farmer Colin Mathieson, a Scottish immigrant and a future winner of the title, and two brothers Binks who were tough guys and good at the hammers and weights but not a threat in the caber.

It was a great trip that had many highlights. We were extraordinarily well treated. We were looked after by a sports PR man called Bob Whitman who had a shock of beautiful white hair which accounted for his nickname 'White Cloud'. Now, White Cloud's usual job was looking after Arnold Palmer, Gary Player,

Jack Nicklaus and the rest of the professional golfers, so he was used to treating his people well – down to the little things. When the defending champion found himself short of a clean shirt for a dinner, White Cloud sent him in a taxi to the shop and gave the driver the money to pay for the shirt. We were toured round the country from Travelodge to Travelodge where we came across three things that were new and delicious: Australian red wine, which until then we hadn't realised could be drinkable; oysters grilled in mornay sauce; and carpetbaggers (stuffed pepper steaks). And we did eat well. White Cloud, who had a budget to feed us, soon ran out of money and I enjoyed him saying, 'Charlie, you guys don't drink that much but by Christ you can eat.' One of the highlights for me was that we were there just as they were finishing the building of the Sydney Opera House. I've always been interested in buildings and for me the Opera House was just stunning. I had been used to saying that modern architects built nothing but ugly boxes, but this was beautiful and so clever, the way the curves of the roofs mirrored the sails of the yachts in Sydney Harbour. Almost as stunning was the reaction of the Australians to the building which was to become their country's icon even above the kangaroo. Not one Aussie had anything but scandalised abuse for their opera house. 'Did you see what they're building there, Charlie? Do you know how much that fuckin thing is costing us? It's ten million dollars.' That must be just about the best £4,000,000 that has ever been spent. I often wonder what they think of it now.

We competed at the Melbourne cricket ground, the Sydney showground and at a public park in the capital, Canberra. There we were near the world's biggest Burns Club, which has a statue of the bard sitting opposite it. They told us that the Canberra Jewish community were building a synagogue next door and that they were going to change the title on our Robbie to Rabbi Burns. I don't suppose I'll ever get back there to see if they did.

The big event was in a park in Geelong, where we caused what we were told were the biggest tailbacks there had ever been as the people came to see the Highland Games. There were really only four of us who were capable of tossing this very big caber, the selection of which Colin Mathieson had supervised. We had all tossed it in practice and I knew I had a good chance. The other three were all stronger than me and I would have no hope if it was too heavy, but if I could toss it I could win it. Disaster struck. The first to throw was Arthur Rowe. He made a mess of it and laid the stick down so awkwardly that it snapped in two. There was no plan B. We'd come halfway round the world for a shambles.

However, we managed to retrieve the situation, somewhat. From a pile of very heavy thin poles which happened to be lying in the corner of the park, we were able to get a stick of about the right length. But it quickly became clear that we had a problem. The original caber had been carefully chosen and prepared so as to be tossable, but only just. The heavy substitute (I think it was eucalyptus) had no taper on it, and we might not be able to toss it at all.

Still, despite what I said about not having any chance if it was a heavy stick, I did manage with my second toss to hit it just right, and while it wasn't a real toss, it was close and just slid away to the side. That was clearly the best attempt until the very last throw of the competition when the Englishman came to take his final throw. It was not a thing of beauty. It was nothing like straight, but it was better than mine. I had to be content with second place. Still, the best caber-tosser ever, Bill Anderson of Bucksburn, was one place behind me, so I had not done myself any injustice.

CHAPTER TEN

George Clark the Highland Games Anti-hero

1958–67

My first close encounter with that great athlete and character, George Clark of Grange, was at Blackford Games in 1958. I was going to a party in Aberdeen afterwards and had, not exactly a date, but something between a hope and an expectation, with Aggie Tosh, one of my fellow students, but I had a problem. I had reached the Games by train and bus and the chances of getting to Aberdeen in time were small. I somehow dared to ask the great man for a lift and he said 'Aye, fairly that, loonie.' We had just left the Games field when Geordie leered over at me and said with the sort of studied charm for which he was famous, 'Ye look like a bugger that could drive?' It was both a question and an invitation to take the wheel. I was delighted. How quickly we got up the road to Aberdeen would now be in my own hands. 'Great! Look out, Aggie Tosh. Here I come.' Or so I thought.

How wrong I was. One mile out of Blackford and speeding towards Perth, just before the Gleneagles railway station, George leaned across me and pointing to the right said, 'Doon here, loon.' It was not a short cut. We were heading down Glendevon and at right-angles to the direction in which I wanted to proceed. Soon we passed the castle which, when it became too dilapidated still to be used for a shepherd's cottage, my parents rented for a while in the 1930s.

My parents had thought it very romantic, living in a baronial tower house. They both wrote. My father went fishing up the tiny burn that runs past and caught delicious brown trout and cooked them for their breakfast. And when my cousin James Mackie was born and the pair became an uncle and aunt, my father made up the following ditty in celebration:

Auntie Jean and Uncle John
Are fond of living all alone.
They never get up till after eleven,
In their castle at Glendevon.

Well, that was all very well for them, but it wasn't a sentimental journey I was on. I wanted to be getting on the road to Aberdeen. Worse and worse! At the foot of the glen I made to turn left for Fife and the road which might have taken us back to Perth, but no. It was, 'Hudd on, min' and furious gestures to turn right along the Hillfoots road towards Stirling.

At Dollar we stopped at a rundown pub. We went into the bar where George said to the barman, 'A treble tae me and a nip tae the laddie.' I suppose as driver I was lucky to get a nip, but it did make it absolutely clear that ours was not a symmetrical relationship. I was the driver and that was it. George disappeared through the back for perhaps half an hour, by which time I was well and truly finished my nip and 'fittin aboot' to get up the road. He reappeared with an intriguing box under his arm. 'Right loon. We'll awa. Mak for Dunfermline.' Dunfermline? Where would it be after that, Edinburgh? Newcastle?

In Dunfermline George carried the box from Dollar into the rundown pub at which we stopped. Again it was, 'A treble tae me and a nip tae the laddie.' And George disappeared into the back shop. I had plenty of time so I passed some of it wondering why, if I was only the driver, and was only allowed one nip at each stop, the great man didn't ask me to carry the boxes in and out of these places? Clearly the boxes were important and I was not.

After only a few minutes in the back shop George was ready once more for the road. Boxes in the car, we set off for Dundee.

I wish I could recall the conversation I had with the great man. For George Clark was a great man in his own way. He was six foot three and seventeen stone in his potestatur. As well as being a Highland Games champion for many years, taking first prizes at Aboyne in his twenties and in his fifties, he was a world champion all-in wrestler who was giving a challenger a doing in the ring in America when the victim's wife came up between the rounds and stabbed George in the back. I used to boast that I had seen the scar, but I cannot now recall that detail. George knew all my family and doubtless sold them the contents of his boxes; he knew all about Highland Games and I had it all to learn, so we would have had plenty to talk about. I just remember one thing, and it was quite interesting. I asked the great man if he declared his winnings from the Games to the taxman. He said yes he did and that he was glad to do it. Otherwise he might have had to reveal his far greater income from the boxes. But apart from that, all I can remember is George, as the trebles began to assert themselves, becoming more and more agitated at how slowly his driver was driving. He could not bear to be stuck behind other vehicles. 'Go on min. Go on. Ye've a clear road aa the wey tae Aiberdeen.' What we were doing driving up Glenfarg I have tried unsuccessfully to remember, but I do remember it clearly. We were stuck behind a slow-moving bus, and it is a very winding and rather beautiful wooded road up to the Bein Inn where, before they built the motorway, the turn-off for Fife used to be. George was furious. 'Go on min. Get by, get by.' By the time the terrified driver was forced past the bus on a not too blind corner, George was beside himself. He screwed down his window and shouted out in a voice that could have been heard all over Pittodrie at a Rangers match, 'Hey min. That's a bus ye're drivin, nae a traivelin shitehoose.' I did not think it was repartee of a high order.

So then it was up to Dundee where, in the Hawkhill, it might even have been in the Marine Bar where years later as a young university lecturer I was to while away many happy lunch hours, we went in for a treble and a nip to the laddie while the manager produced the biggest television set I had ever seen. The screen must have been thirty inches and, remember, in 1958 a Murphy's Ten Inch was a big TV set. 'God be here, min Albert. That's nae a TV,' George exclaimed, 'it's the Gaumont Picter Hoose ye're sellin me noo.' That called for a second unbalanced round before we loaded our picter hoose and set out for home . . . or at least, so I thought.

George was even more anxious to get his driver to make good time, and when we approached a set of traffic lights in the middle of Dundee he just wouldn't hear of me stopping despite the fact that they were red. In fear for my life or even worse I managed to jowk through the lights to some tooting of horns but no great danger. The pubs were just spilling out, which meant that it was half past nine already and there were plenty of witnesses. Somebody shouted at us. It sounded like 'You've got a puncture.' So I stopped and George got out. And made to inspect the wheels. 'Ye should be in the jile,' one of the bystanders opined very fairly, I thought. To that George Clark responded with a piece of the repartee for which he was famous. 'Fit's a dae wi you, loonie. Is't yer erse?' By the time we got to Gourdon, the front door of the pub had long been shut – though there was a good-going lock-in in progress. Again it was trebles, and nips for the laddie, and more boxes – one delivered and one uplifted, before we set sail once more for Aberdeen. It would be midnight by the time we got there but the party would still be going, and who knew, Aggie Tosh might not be away home yet.

Though still billed as 'George Clark of Grange' in Banffshire, George was now living in Torphins, so when we got to Stonehaven he wanted to take The Slug Road over the hill to Deeside. Without warning he said, 'Well loon, this is far I leave ye.' Wait a minute. Had he forgotten that I was driving? I could have insisted, couldn't I? Wrong again. It didn't even cross my mind that I would challenge the great man to be reasonable. My quick run home was in ruins. But George wasn't quite as bad as that seemed. He bade me stop the car while he got out and stood in the middle of the road. It didn't even occur to me that I could have driven off and left him to it. After all, that had been his apparent intention for me.

The first piece of traffic we saw was a van, driven by an oldish man who had been on commercial business at one of the shows that day. 'Ye'll tak this loon tae Aiberdeen,' said George Clark without even a 'fit like' to start the conversation. 'Oh certainly, Mr Clark. It'll be a pleasure, Mr Clark. Any thing to oblige you, Mr Clark,' was the nervous reply. Although it was out of his way he took me right to the flat in Ferryhill where, sadly, Aggie Tosh was busy. What might have been if Geordie had taken me right up the road – might have been.

So what was in these boxes? Disappointingly perhaps, I think it was television sets. They were subject to resale price maintenance at that time and George did deal in cheap sets. Also there may have been a question of purchase tax, which was very high on luxuries like television sets. Buyers didn't get a written guarantee with their purchase. They just had George's personal guarantee. Bill Anderson, the great Highland Games champion, had one of Auld Clarkie's TVs. If anything went wrong George came and took it away and left him a replacement. It would have been tempting to think of the great man as one of the early cocaine dealers – but no, I think it was just tellies.

George Clark was well known on the circuit for his mischief. He arrived at Lochearnhead Games with a friend whom he persuaded to tell Jimmy Miller, the commentator, that George Clark had died. Jimmy immediately charged into eulogy mode and gave a tribute to the 'legendary George Clark, the all-time great athlete' and then asked for two minutes' silence. During the respectful silence George strode onto the field and tapped Jimmy on the shoulder, uttering the words 'Cauld day, Jimmy.' Jimmy turned round and nearly fainted but recovered to utter the words, 'Clark – I never liked you. You were aye a devious bugger.' Unfortunately the microphone was still on, so Jimmy's glowing tribute was somewhat devalued.

I am loath to move on from George Clark, for he was a one-off among one-offs. He was one of those people who command a room. I have known many to whom everyone would turn in case they missed something. George was like that, but he had the added dimension of menace. You were never quite sure that you would survive an encounter with him. Many people never recovered from the shock of the first encounter with Auld Clark, as he had become by the time I was a fellow competitor at the Games. The first time that I became a real competitor to him was at Strathpeffer Games, where I managed to beat him at the shot putting. After the heavies were finished he said to me in a very gruff way, 'Are ye gaun ower tae the tent, min?' I said no. I was still involved in the light events at that time and I wanted to stay and watch the pole-vaulters. It was a social gaffe on my part. George had not been asking what I was going to be doing next. By way of encourage-ment he was offering me a drink in the tent. I had rejected a sign of acceptance that I had arrived on the heavyweight circuit.

It did dawn on me that I had made a mistake and I had an early opportunity to put it right. As luck would have it, Auld Geordie's car was parked next to ours and when we were about ready to go the great man appeared and was also packing. Now, it so happened that I had a bottle of Glenmorangie, the excellent Ross-shire single malt whisky, and there was maybe a quarter of it left. I handed it over and said, with a look which I hoped would be recognised as an apology, 'A drink, George?' He took the bottle without a word, put it up to his great lips and scoffed the lot in a oner. He handed back the empty bottle with the words, 'Worst whisky that ever I tasted.' I was disappointed, but my poor lady wife was astonished. The family of a banker in Broughty Ferry didn't get many examples of manners like that. Indeed, neither did the family of a peasant farmer in Buchan.

And still I can't bear to leave Auld Geordie. There is so much to tell. It is not all good, but I shall tell the best of what sticks with me. The first is an example of his sense of humour. There are not that many, so I cherish this. It was at Fort William Games in 1964. Geordie had, I discovered later, spent a busy night on the tele-phone the night before, telling his opponents that Bill Anderson, by far the outstanding heavyweight of those days, and his nearest rival, Sandy Sutherland of Alness, were both going to Fort William, so they would be best to divide their attentions between Aberlour Games farther north and Thornton in Fife. That would leave a fine easy day for himself at Fort William. With those two stars taking all the firsts and seconds, Geordie would have a field day among the lesser prizes. And his plan was even more cunning than that. Bill was indeed going to be there, but it was just a lie about Sutherland, who he well knew was going to Aberlour.

It worked well for Auld Clark in the end, and much better than he could have hoped really, because he managed to beat Bill Anderson in the weight-over-the-bar. It was the only time I ever saw it, but Bill failed to clear a low height and even I beat him that day. But there was much anxiety before Clark saw the wages of his deception. It was a really foul day and the start was delayed in the hope that the rain would clear, or at least ease off a bit. This was a nuisance for me, but Auld Clark was clearly agitated. He paced up and down and looked out of the tent every few minutes and muttered to himself. I only realised later what he was worried about. Aberlour was only a good hour away. The lads would have discovered his deception by now and every minute of delay made it more likely that the dreaded Sandy Gray, who could beat him easily in most events, or Bob Aitken, who would beat him in all but the caber and weight-over-the-bar, would appear on the field. Eventually, peering out between the wall and the roof of the tent, he exclaimed, 'I was here thirty year ago and it was raining then and I have been here every year since and it has aye been raining.' And then, squinting up at Ben Nevis, 'In fact they'll never get a richt day here till they shift that fuckin heap.' So much for Britain's highest mountain.

I suppose misdirecting his opponents would be excused as gamesmanship. But Auld Clark was also a spectacularly successful cheat. My favourite example is the time he took the fourth prize in throwing the light hammer at Oldmeldrum Sports in 1967.

By this time Geordie was no longer competing, except on the odd occasion when he could so arrange it that most of his rivals didn't turn up. He was there in the strategic role of judge. Now, the ground at Oldmeldrum doubles as the football ground of Juvenile Football stars Oldmeldrum AFC and the ground was bare of grass and very hard because of the combination of many football boots and no rain. Each throw of the hammer was followed by much searching the ground to find exactly where the hammer had fallen. It so happened that while Bill Anderson had had his usual prodigious throw and was easily first, and Bob Aitken and I had had reasonable throws for second and third, there was an unusually long distance before a fourth-placed throw could be found. While everyone was looking for the mark of one throw, Auld Clark picked up the hammer and, in a gesture very common in hammer-throwing, gave it a mighty dunt on the ground, supposedly to shake off any mud, though there was no possibility of mud on that dry pitch, and then said, 'Ach, I'll just tak a throw.' As he was the judge as well as being George Clark of Grange, who was there to deny him? With much huffing and puffing he then took a canny-like throw, not bad, but hardly a prize-winner. Everyone looked hard to see where it had landed, though what did it matter whether he was fifth or sixth. 'Na, na, boys,' the judge declared, 'I can see the mark frae here.' And he marched up past the searchers to where he had made his fierce dunt before his throw, and there was a clear hole, and in fourth place. And who was to argue with the judge?

Earlier he had beaten the great all-rounder Jay Scott of Inchmurren by a Machiavellian piece of cheating. The young Scott was just too strong for George by this time and was lying a clear first in throwing the 28 lb weight. Now, the Games park at Ceres is very narrow indeed. Unless you saw it on a Games day you would just think it a useless strip of grass by the river, big enough for a few picnickers but little else. Somehow they hold a Games there that has been popular since it was first celebrated by the victorious Scots on their way to their homes from Bannockburn. With his last throw, Clark let out a yell as he was winding up, let it go and made a great show of how he had been put off by one of the crowd who had got a bit too close to his line of throw. Meanwhile the weight flew off and landed in the burn. Clark marched furiously after it, waded in and returned to retake his last throw. This time all was well, and with a mighty heave the old man won by a narrow margin. Now, all of that is true and I dearly hope the next bit is too. Jay Scott swore that Geordie had concealed a similar but lighter weight in the burn before the event. He then waded in, not to get the competition weight but one that was easier to throw.

And George was at it again when, at the very ripe age of fifty-two, he won the first prize at the Great Aboyne Games in the strongest event of all, throwing for distance the fifty-six-pound weight by ring. Surely even Clark couldn't hoodwink the judges at Aboyne? Well, he did, and it was possible in those days, because at Aboyne they had a very odd way of measuring throws. Instead of measuring to the nearest mark on the grass, they stuck a peg in the middle of the hole and measured to the top of the peg. Now Ek Wallace and Clark were well out in front, and, by applying a string from the centre of the trig, the judges and all the competitors were quite sure that Wallace was in front by almost an inch. But see the master swick in action.

The throwing events paused so that all eyes could turn to the final of the 100 yards race —all eyes except for three pairs: Geordie's and the witnesses' (Bob Aitken and Charlie Allan), who were sitting on a caber and just watching. As the excitement of the race intensified, we saw Geordie take a stroll through among the pegs, and without seeming to even look at them, he nudged his peg so that it sloped slightly forward and he nudged Wallace's peg so that it sloped ever so slightly back. When the learned judges came to measure the throws they found that Clark was first by an inch instead of being an inch behind. In sport as in life, there is more to winning than being the best – though in this case Auld Clark was the best – at cheating. It was a shame for Ek Wallace. He competed well at Aboyne for many years but he never did get a first prize, though on that occasion he did deserve it. I think that was the last time Aboyne measured in that eccentric fashion.

I liked Geordie Clark. He was not boring. And if much of his behaviour was unethical, he was a rich part of life's tapestry.

CHAPTER ELEVEN

Have Kilt – Will Travel

1962–71

In the early 1960s there was an exciting development for the heavyweights. We started to be asked to throw things overseas – the lucky ones among us became 'internationals'. Jimmy Miller, the great blether who commentated at many of the Games in Fife and the Southern Highlands, like Inveraray Castle (the home in Argyll of the Campbells) and Luss on Loch Lomondside, even made me blush when I ran onto the field, late as not unusual, sporting a most unusual tan on my return from three weeks' competing in the Bahamas. Jimmy, in a string of laudatory nonsense, described me as, 'Another of our globe-trotting athletes'. It was the high point of a very poor day in the field. The globe-trotter was exhausted after his flight from Bermuda and only won one second prize out of nine competitions.

The expansion of the world of the heavy athlete was really just another of those things which had been before but which had been halted by the war. In the '30s George Clark had competed all over the place, mainly as an all-in wrestler, but his predecessors A. A. Cameron and Donald Dinnie had travelled the world with strong-man acts, including tossing the caber, throwing the fifty-six-pound weight and the rest. New overseas interest grew out of a stage show at the Kelvin Hall fronted by Archie McCulloch, the Glasgow music hall personality. It was the brainchild of David Webster, and was really a cross between a stage show and a Games, with Highland dancers doing the Highland Fling and the Sword Dance.

There was some fine singing by someone reckoned to be the up-and-coming challenger in the Kenneth McKellar school of Scottish singing. The three heroes were Jay Scott, off whom age and injury were taking the shine, but as far as the Glasgow public was concerned, the star. He was a very bonny man and had married Fay Lenore, the music hall singer and pantomime boy, in what the yellow press had called the wedding of the year. Bill Anderson, now the unchallenged champion, was there of course, and Charlie Allan was there because he laughed a lot, didn't cost much, as he was a junior lecturer at Glasgow University and had nothing better to do. We couldn't throw the hammer in the arena, and putting the stone was reckoned to be too dour for a stage show, but we did the caber, the weight-over-the-bar and Jay and I did the high jump. Jay's style of jumping meant he landed on his feet, but my elegant straddle meant that my shoulder and elbows came clattering down on the boards from five foot six or more. In addition, Jay and I gave a demonstration of 'tilting the bucket'. That is where one sits in a wheelbarrow while the other charges it at a contraption in which there is a small hole through which the competitor in the barrow has to poke a long pole. If he misses the hole a bucket of water falls on the competitors. Guess who got the job of sitting in the barrow? And it was even less fair than it sounds. Jay soon learned the exact speed necessary to ensure that all the water fell on the occupant of the barrow and none on the driver. I got very good at getting the long pole neatly through the hole. But that, according to Archie McCulloch, was no fun. I was ordered to fail at least once and ensure the crowd a laugh and myself a soaking.

That wasn't much like the Highland Games, and I soon got another reminder that we were not in athletics but in entertainment. The challenge in 'throwing the weight for height' is to throw four stones over a bar which is moved up and up until all but the winner have failed. It is like the high jump, but you don't have to crash land on a hard arena floor. Now, I had some difficulty, not in throwing the weight high enough, but in aiming right – the weight often came crashing down on top of the bar instead of sneaking up one side and down the other. On the first and second nights at the Kelvin Hall arena, Jay Scott and Bill Anderson were by far the best, making the feat look boringly easy, while the junior lecturer, the 'local boy' not half a mile from his day job, struggled. It was then that I learned something about showmanship. They didn't eliminate me even if I didn't get the weight over, and the bar was raised higher and higher until, at the respectable height of twelve feet six inches, they came to the top of the apparatus. It had been made for the pole vault and no vaulter in Scotland could beat that in those days of steel poles.

But this was the weight over the bar, and while twelve feet six inches was more than I had ever done the other two could throw the fifty-six-pound weight over fourteen feet. It seemed clear that Jay and Bill would be first equal and I would be a long way off in third place. The other two cleared the top of the posts with some ease and had the cheek to shake hands. But what about the laddie? Well, goodness knows how, but I cleared it to thunderous applause. The Harlem Globetrotters could not have done it better. Not only that, but I did it again the second night.

On the third night of our two-week sell-out run, however, Fiona was in the audience. I was determined to do much better. I made sure I was properly warmed up and mentally keen for all my early throws, which didn't sail over like those of Anderson and Scott, but I did clear all the heights at the first attempt. I was first equal again. Fiona was proud. But Archie McCulloch was furious. I was summoned to a meeting in his office. It was a boring spectacle with the three of us just chucking it over every time. I was ordered to fail at the lower heights and succeed only with my final effort at the top height. I did my best, but I was at the limit of my range, and it was not easy and I often failed to clear the final hurdle and gain my burst of applause. It only occurs to me while writing this forty-eight years later that what they should have done was to get the other two to take turns at a last-gasp success; they after all had a big margin for error. Jay, at least, would have been pleased to fake it in the interests of applause. Anyway, Archie McCulloch, who had regarded me as a star, never showed me the same respect after that.

Well, well. I much preferred the genuine competition of the

Highland Games. But there is no doubt that the huge kilties showing their dour strength and the delicate dancers strutting their stuff while the pipers filled the air with the sounds of old Scotland had a market, and soon we were invited to provide photo opportunities in exotic places. Quite the best part of this development was that it brought the next generation from Ardo's hill into contact with Wild Ian Campbell – but more of Ian a little later, as he wasn't on my first trip abroad.

In 1964 I went to Nassau in the Bahamas to put on a Highland Games show for the entertainment of the tourists who flocked to the Caribbean islands and landed at least twice a week from giant cruise ships. Now, you could be forgiven for thinking that those who went there would be looking for steel bands and limbo dancers, but the Bahamian equivalent of Archie McCulloch had been persuaded that there would also be a market for hairy knees, kilts and bagpipes. Again the team was Jay Scott, Bill Anderson and myself. We were the heavy athletes. But this was an ambitious show. The Bahamians had been sold a pipe band from the Glasgow University Officers Training Corps, and a team of four amateur boxers. The event was to be sponsored by Ballantine's whisky and their managing director's daughter and her pal had taken several lessons in Highland dancing so that they could showcase Scottish dance, and for some reason the Scottish go-kart champion was also there to take on all comers.

It all seemed a bit improbable to me but they were offering me £80 a week all found and the chance to see Nassau, where weeks earlier Harold Macmillan had met Jack Kennedy to decide on a policy for the containment of the Soviet hordes. It was three times my salary at the University of St Andrews and I was going to get that as well – for eight weeks! That was a long time to be away from Fiona and the children, but it was exciting.

When we arrived at Nassau it began to seem a bit less improbable. We got a great reception at the airport from a respectable and excitable crowd. The publicity at least had been well handled. One local pushed his way to the front and said 'I gat to see that Jay Scott – he's twenty feet tall.' We saw why on the way from the airport to our hotel. Just outside the airport there was a huge cut-out of a kilted Jay Scott which might have been twenty feet tall and was certainly much more than his six foot three. Then we saw where our shows should have been held. On a hill near the centre of Nassau is a castle, Fort Nassau, which had been built in the days of the slave trade by the British to protect their colony. It was floodlit when we first saw it and it reminded us immediately of Edinburgh Castle, and the cricket field below was absolutely perfect for a Highland Games. A couple of miles down the road and we were walking through the grounds of our old colonial hotel where the band which played about fifteen feet up a giant fig tree each night was knocking six 'Yella Birds' out of their steel drums while the blue of the swimming pool shimmered temptingly at us between the trees. That was the good bit. The rest was an unutterable shambles.

The venue was not below the castle but twelve miles out in the desert on an abandoned race-track. A temporary arena had been erected with seating for 2,000 and changing rooms. To keep out the hoards of freeloaders who would otherwise have dodged paying to see us there was a fence of bare plywood boarding about eight feet high. The pipe band did a grand job in painting Disneyland Scottish hills and skies of blue on the clapperboard, but that only helped by covering the bare wood. The organiser had been too busy, it was said, renewing his acquaintance with one of the sponsors' wives to do anything about selling the attraction to the cruise ships, so we didn't see a single cruiser. So much for the big advertising campaign that was supposed to lead to a huge influx to the Bahamas from New York and Miami.

On the opening night we were appalled. A few local Scots were there and that was it. The announcer, freed from his amorous duties for the night, made a portentous announcement: 'As there is only a relatively small crowd tonight we would like you to feel free to take any seat you want even if you have only paid for the cheap bleachers.' That night's crowd was by some way the biggest we had. On the second night I was making the boys laugh by mimicking the previous announcement with 'As there is only a relatively small man in the crowd tonight we would like to invite him to join us in the royal box.' It was all over in three weeks. We got paid though, and I learned to snorkel at astonishing coral reefs where I was underflown by a stingray that, with the magnifying effect of the clear water, looked like it was ten feet across. It was beautiful, but I made it to the boat in double-quick time. Better than that, I became known to Jimmy Miller as a globe-trotting athlete.

Then David Webster managed to sell the idea of the hairy knees and bagpipes to the McVitie's biscuit people. That meant visits to two international trade fairs, one in Tokyo and one in San Francisco. The team was Bill Anderson, Charlie Simpson the Caithness policeman, myself and Dave Prowse, an Englishman, who had several physical claims to fame. He was six foot seven tall, had been sixth in the Mr Universe competition, and his size fourteen misshapen feet had played the part of the villain's feet in *Count Dracula*. That was not a difficult part as those feet only had a swim-on part, appearing at the end out of the vat of acid in which the villain died. There were four dancers: Billy Forsyth, the then world champion from Stirling, Dougie Duncan from Bucksburn, and Arleen Stewart and Margaret Murphy, both from Aberdeen. The piper was Ian Blair from the Aberdeen Police Pipe Band. That was the Highland Games contingent, but then there were the showbiz add-ons. Wild Ian Campbell, the first TV voice of professional wrestling and himself a European heavyweight champion, was there. He was the ad man's ideal Highland Scot, huge, with a thick black beard. And there was Clayton Thomson, a much more skilful wrestler but much smaller. And Graham Brown, 'the Highland Hercules' from Lowland Arbroath, was also there – a small man who had built up a powerful body in the gymnasium. He was the strongman. Graham had a number of the usual tricks, like allowing people to break boulders on his chest, and blowing up hot water bottles. But he also had one extraordinary feat of daring and self-control rather than strength. He could stick a six-inch nail up his nose, right up to the hilt, and for extra effect he used a little hammer to tap it in the last inch or so while sensitive members of the audience were quietly sick. Don't try it up your nose, but hold a six-inch nail against the side of your head and you'll find that it should be just about ready to break through the scalp when hammered fully in. Graham also had a steel rod about twelve feet long which he used to bend by leaning the razor-sharp blade against his throat and shoving. Of course, Graham's tricks were nothing to do with Highland Games and not that much to do with strength. The six-inch nail, for example, didn't go up through his brain. He put it up his nose all right, but from there it went down his throat because of the clever way he positioned his neck. It still took a lot of self-control not to sneeze this foreign body out of his nose and retch it out of his throat, and the blood he often drew was genuine, but there was no need for a neurosurgeon. And the steel pole was not bent by pushing against the razor blade with his bare throat. That would certainly have killed him. He gripped the blade between his chin and his chest and bent the steel rod by leaning on it.

The arrival of this lot at the airports of the world was nothing like what happened when we athletes shuffled onto the stage at Nassau and posed modestly for a photograph. When we got off the plane at Tokyo the dancers dashed around looking for something to climb and pose on, and the strongmen looked for something to lift. Anything would do. A trolley-full of baggage at Heathrow, a giant spinning globe in San Francisco, a car in Tokyo - anything that looked heavy but wasn't too heavy and would make a good picture.

The idea was to put on Highland Games in a public park. In San Francisco we were in the park beside the Golden Gate Bridge. It had been laid out by the Scotsman John Muir, the father of the National Parks movement in the States who established the famous Yellowstone Park. Between the athletic events the big men in their skirts would parade around town turning heads to read the tee-shirts with McVitie's written on them. We athletes hated those tee-shirts. It was all right, we thought, as part of showbusiness, but we were athletes and didn't like wearing 'sandwich boards' as we called them. We also got very fed up of all the ladies who came up to us everywhere and asked us a lot of daft questions before getting round to the daftest of all: 'What is worn under the kilt?' Sometimes, just to please them, we would reply 'Nothing is worn; it is all in perfect working order,' but usually we just smiled good-humouredly, for they were just being friendly as Americans usually are. But it did get a bit wearing. In Fisherman's Wharf in San Francisco, Wild Ian Campbell astonished us all when the umpteenth lady of a certain age came bustling up to us and said, 'Say, are you the guys who throw the pole thing? I saw you on television.'

'No, missus,' said Campbell in his strong Fife accent and double bass voice that sounded like a cement mixer. Pointing to the McVitie's sign on his chest he delivered the unforgettable line, 'We're the McVitus dancers.' It wasn't very politically correct, but then Ian wasn't that.

In Tokyo in 1969 we put on a Highland Games in Toshimaen Park, an outdoor stadium which included a leisure centre with the scariest rollercoaster I had ever see or have ever seen since. We were a culture shock to the Japanese, with our size and our kilts, but they were a culture shock to us too. In those days they all dressed alike. The old people in traditional dress, but the young thrusters, which included just about everyone in Tokyo, wore black suits and white shirts and really, to our eyes, with their clear skins and beautiful jet black hair, they were identical. They were very polite and could never understand what we were laughing at. With Campbell it was just as well. We had a very worried little host, whose unenviable task was to get his charges to wherever they were going on time, all present and correct. We called him 'Mr Hurry-Hurry-Bus-Leave-Half-Past-Eight'. Campbell managed to get this poor harassed man inveigled into a drinking session at which he showed him the old Scottish toast. He got the wee man to lock his spindly arms with his huge hams and repeat with him the immortal salutation, 'Bollocks'. Now, another part in the saga of that trade mission to Japan was that we also had a minor member of the British royalty with us, there to add even more dignity to the occasion. At our biggest event we were all given a knock-doon to Prince William of Gloucester. That all went fine until it fell to Mr Hurry-Hurry-Bus-Leave-Half-Past-Eight to make a toast to the royal visitor as he left the field . . .

That was deliberate and mischievous, but there were great cultural difficulties to be overcome. The starkest of them involved our interpreters. We had one each, university students brushing up on their English and earning a few yen in their holidays. They really needed to brush up if their English was to be any use to them in Aberdeenshire at least. Their Japanese was perfect but, while they could say quite a few words in English, they had hardly any understanding of what English speakers were saying. I was anxious to learn as much as possible and was always bombarding my interpreter with questions, so, to protect the girls, I was given the best one, who was an advanced student of English. But still, communication was difficult and often non-existent. My best illustration is when we were driving through Tokyo and passed a fantastic pagoda. It looked to me like a religious building of some kind, perhaps a Shinto shrine to some great Japanese warlord of the fifth century. It towered above everything like an enormous fifteen-tiered wedding cake. It was rather as though they had built one storey and put on its nice curvy roof, then decided they would like another storey and built it on top. Then another and another. It stood maybe twice the height of any of the modern buildings we'd seen and was perhaps ten times as high as the traditional buildings which we saw all over the place as we travelled around.

'What is that?' I asked my guide, pointing. She looked quite bewildered. Then she had a flash of insight. The worried look gave way to a knowing smile. She knew what the white man wanted to know and she could help after all.

'Ah so,' she said. 'Is o rady is coming flom frowallangeing crasses.'(It is an old lady who is coming from her flower-arranging classes.) Now I was the one who was bewildered. I had heard the words, but what did they mean? After staring at the scene for a few moments I at last saw a little old lady in traditional peasant black, hurrying along with her tiny steps, carrying a big bunch of flowers. My interpreter never realised that it was the huge building in the

131

background that had aroused my interest. This was not a meeting of minds.

Of course we knew almost nothing about Japan or the Japanese except that they did the free world a great service by bombing the United States into the Second World War in 1941, and that they were hard on their prisoners. But we had heard of geisha girls. We had heard tales of weary travellers being so royally entertained that the geisha girls who looked after their feeding and watering were also expected to do all in their power to make them sleep as well as possible. It was a relief to find out that, although we were well treated, we were not important enough to be embarrassed at bedtime. We did see an occasional hostess who was a geisha, but a shoulder massage, available at no extra charge in any up-market bar, was as far as we were able to see. In the dives we went to the offer of a hot towel to wash one's hands on arrival, a novelty to Scots in 1968, was as far as personal service went. However, we were advised that we could not leave Tokyo without experiencing a bath house; indeed it would be an insult to our hosts if we left without doing so. In the interest of hygiene, and out of politeness, most of us went along.

After paying 1,500 yen (about £1.50 in today's money) through a little wire-mesh hatch we sat in a little room that reminded me of where you wait for your order in the lowest takeaway restaurants. After perhaps ten minutes a little Japanese woman who was neither old nor young and clad for the gymnasium came in and silently bade me follow her. I don't remember her having a name, but as I try to relive the experience for you I'll need to give her one. Let us call her Ginza.

Ginza took my clothes as though, in that land of silks, she had unfrocked a thousand pipers and had seen many men who wore skirts made of thirty feet of thick pleated wool. She was not at all uneasy about the fact that her client was armed with a fancy knife which he concealed in his sock. She folded my modesty neatly and put it away in a cupboard. It was all humdrum. When she had got me stripped for action, Ginza used her only words of English. 'Beega peenis,' she said, with as much feeling as a supermarket shopping-bag filler might have said, 'Fine day.' Perhaps she didn't even know what it meant, but had just noticed that when she said it and waved a hand at the midriff, the white men among her clients looked pleased.

First Ginza put me in the steam washer. This was a series of wooden slats under which were hot stones. Then she put a layer of slats on top of me so that I was loosely sealed. When she poured water on the stones a delicious cloud of steam came up and started cooking me and sweating the dirt of ages out of my everything. It wasn't so delicious when the heat was intensified. Then a sort of competition developed in which Ginza held on the steam and I pretended that I wasn't bothered. Eventually, I could stand it no longer and burst from my scalding steamy trap. Had I appreciated what was to come I'd have been out of there a lot sooner, I can tell you.

The next step was for the client to sit on a three-pronged stool perhaps six inches high. That left him in a ludicrous knees-in-the-air position. I felt very vulnerable – which I was. But that was exactly what was required, for Ginza was not going to give me a cat's lick. It was a point of professional pride with her to scrub out every nook and cranny. I got clean in places I hadn't until then noticed. She was into everything and under everything. She pulled back everything to make sure she could get access to everything else. She particularly liked my ears, where I thought at one stage she was trying to get her cloth in at one side and out the other so that she could give me a pull-through. When the little inflatable thing at the front started to raise his ugly head, which was soon and often, she went to this big trough and produced a bucket of water which wasn't quite boiling and threw it over him. That quietened him for minutes at a time. Several times she got her client thoroughly soaped from top to toe and then splashed him down, for all the world like Jimmy Kelman sweeling the byre at Little Ardo.

Then the client was dried just as thoroughly and oiled, again everywhere, for his massage. I've had quite a lot of massage over the years for all my sporting aches and sprains and I have to say in that Ginza's massage was disappointing. She was full of tricks, like clicking her fingers as she pummelled, which might have given a feeling that her pummelling was getting somewhere, but she just hadn't the strength to make any impression on a large, fit Highland Games athlete.

The whole bathhouse experience was so matter-of-fact that, although it was of course very sensuous, I found it quite non-erotic. My main feeling about it was that it was very funny. When one of our number got to the bit where he had to lower his twenty stone onto the little three-pronged stool the three legs just sprayed out and he, who had already had his dignity hung up neatly for him in the cupboard, landed flat out on the bathhouse floor in a little heap of kindling. Perhaps the funniest for me was being far from home, naked, with a little girl running barefoot up and down my back making sure my spine was all in place. But not every one of us was so unaffected. One of our number was aroused sufficiently to ask if there were any special extras available. Indeed there were, but they would cost. His little girl produced a piece of paper with two figures written on it, 2,000 yen (about £2) and 3,000 yen. Well, he thought he knew what the 2,000 yen would buy him. But he was far from home and he might never pass that way again, so he felt he deserved something special. Damning the expense he pointed to the 3,000-yen sign. Immediately and without ceremony, his masseuse fell upon him manually. That had been pleasant enough but our friend was puzzled. If that was all you got for the 3,000 what could have been on offer for 2,000? The only thing he could think of, and he was very proud of the fact that it had been possible, was that the de luxe service was performed with both hands.

A trip to the trade fair in Tokyo was quite something in those days. It was exciting, it was interesting and it was memorable. Tokyo was the least cosmopolitan place I've ever visited. We just never saw another white person. But for me it was memorable in one very unpleasant way. On the way home, I got a glimpse of what madness could be like, and I got to understand how sleep deprivation might be used as a torture. It culminated in my refusal to leave the plane at San Francisco and being made an offer I could not refuse and a choice I could not dodge. Looking down on my frame, recumbent in the foetal position across three seats, an enormous and immaculately uniformed black policeman said, 'Now Mr Charles M. Allan, do you wanna come with me to prison or would you prefer to go with these men in white coats to the hospital?'

It was all down to something called circadian rhythm, now well understood as jet lag. Much of a man's behaviour is determined by his body clock. That's what wakes you at roughly the same time each morning irrespective of when you got to bed and makes you feel tired at bedtime even if you have had a lazy day. When you move between time zones it takes a while for your body clock to change. The worst thing you can do is travel halfway round the world so that night becomes day. And that was just what we had done in flying over the North Pole to Japan. What had been night was now day. But if I couldn't sleep all night that didn't mean I could sleep all day. There was so much excitement rushing at us every day that we didn't have time to feel tired. But when it came time for bed my body clock was telling me it was time to get up and where was my breakfast? I didn't sleep at all in my ten days in Japan, and although I did better than ever before at the throwing and much enjoyed getting piled into the public relations part of the job – diving into crowds and shaking every hand offered – I started to get things wrong.

Just when my circadian rhythm might have been catching up and I started falling asleep briefly at odd times, we were off again, and with another twist of the circadian screw we left Japan and landed at Hawaii. The misunderstandings were always that. It wasn't hallucinations. What I imagined did happen, I just didn't understand it. The example I can remember the most clearly was in a late-night shop in Honolulu where I asked the assistant for disposable razors. He looked me straight in the eye and without a smile or a word thrust his arm at me with a finger pointing straight down. I looked where he was pointing but all I could see was bars of chocolate. I was scared. If disposable razors looked to me like Hershey bars, I was in trouble. Then there was a very loud throbbing buzz pulsing in my ears. I could have panicked but I just stood there – miserable. After a long time, which was probably twenty seconds, I realised that I was standing below the very noisy air conditioner. All I had to do to get rid of the throbbing was to move over a bit. And then I saw, down on the shelf below the Hershey bars, the disposable razors.

It was not understood by anyone in our party. We didn't do much jetting, so jetlag was not an important issue. But it was humdrum to the doctors at San Francisco airport. They whipped me off to Burlingame Hospital, with which I was familiar on account of watching Perry Mason each week. There they gave me a jab which so overwhelmed my body clock that I slept for forty-eight hours. I woke refreshed, very nervous and shaken, but with all my senses restored. Sadly, I had missed my flight home, which was very hard on Fiona who, instead of the wanderer's triumphant return, had to endure a phonecall from David Webster saying he didn't know when I would be home, but I was perfectly all right and not to worry. Some hope.

If that was the lowlight of our trip to Japan, there were a number of other highlights. One was when Graham Brown was bending the steel bar in Tokyo someone got a marvellous photograph of Princess Chi Chi Bo in traditional Japanese dress peeping terrified from behind her fan. And then there was the time the future farmer of Little Ardo attacked the Pipes and Drums of the Royal Scots with a missile. It was at the Toshimaen Park in Tokyo and we were busy at the heavy athletics. I was enjoying the thing best because I had twigged that this oriental crowd had a totally different way of looking at sports. They didn't realise that a throw was finished until, just as in Sumo wrestling for example, the athlete bows. So there was Bill Anderson, throwing everything 10 per cent farther than anyone else and being met with bewildered silence. But my efforts, which won most of the second prizes, were greeted by rapturous applause after I twigged that what you had to do was to bow and show you had finished. So I would throw, turn to the royal box and bow, cueing applause from crowds of up to 22,000 every time. It took the lads a while to cotton on.

Now, there was one event only in which I had any hope of beating the great Bill, and that was the caber. At 25 feet it was the longest we had ever tossed, though far from being untossable as it was light, perhaps one hundredweight.

It is always important with the caber to get properly set for the toss, but with a long, light pole it is important to combine that balance with as much speed as possible to force it over the top. Not used to this length of stick, I found it difficult to get the balance just right, so I sped on and on, much farther than I had intended, determined to get it just right and truly earn my next round of applause. The Pipes and Drums were marching right across my line of fire, though far beyond my usual range. Everyone in Toshimaen Park except the engrossed caber-tosser could see the danger. On I sped, nearer and nearer to the band until I was properly set, and at full speed I dug in my heels and pulled it up and over. It was a pretty good toss and just about straight. But with the extra length of my run and the extra length of the caber it came crashing down amid the Pipes and Drums. There were screams from the crowd and an agitated 'Watch yersels!' from the athlete. But we needn't have worried. It was a perfect example of army discipline. The pipers had been watching and without stopping playing, the row in the firing line simply marked time for two steps, watched the caber thumping into the ground at their feet and stepped over it without breaking their step or their melody.

And my favourite highlight of all, and perhaps for all time, was surely in Hawaii on the way back. Again it involved the wit of Wild Ian Campbell. The local weightlifting club in Honolulu invited us to go to their gymnasium and train with them. This we did, and it was a congenial occasion enjoyed by everyone. We had a good workout, which was very much needed as there was a great deal to work out of our systems after two weeks of Japanese hospitality. They had a lot of good lifters, but the star was a Polynesian man who, at the Olympic Games at Tokyo eight years previously, had won the eight-stone gold medal. After the training we had a shower, of course, and the great Ian was showering next to the wee champion. They were quite a pair, with Ian wild in his big black beard and almost three times the weight of the clean-shaven wee man. Suddenly the Olympic champion gave a little shriek of amusement and, pointing at Wild Ian's manhood, said, 'He, he, he, look at you, big, big man, little, little prick. Look at me, little, little man but look at my prick,' pointing at a thing like a foal's foot that hung down from his lower abdomen. Campbell was magnificent. He pulled himself up to his full six foot four, stuck out his enormous chest even farther and, towering above the little gold medallist, delivered himself of the immortal line, 'In Scotland we breed men, not pricks.'

CHAPTER TWELVE

Back on the Farm the Heroes Retire

1966

What an awfu feel was my auld man,
What an awfu feel was he.
He aye preferred the easy wye
He wasna nane like me.
He wadna buy nae stock nor gear and it's a sad admission.
Rent oot ma grun for the quiet life was the heicht o
   his ambition.

The cows, who had done a great job keeping the cash flowing at Little Ardo during all of my father's time on the place, started to do less well about 1962 – half a dozen years into his illness. How much was due to the fact that Kelman, our excellent cattleman, took semi-retirement with a small herd of Jersey cows on Deeside, I don't know, but cattlemen who were good enough were not easy to find. The set-up at Little Ardo wasn't attractive. Keeping cows in a byre all winter and tying them up twice a day for milking even in summer was hard work, and while one man could handle a byre of forty cows, there wasn't much scope for expansion. Herd sizes were growing. Milking parlours were the way forward, and my father thought he didn't have enough acres to increase the herd enough to justify the expense of conversion. With twice as big a herd, Little Ardo would have been able to pay bigger wages. At any rate, we never got another top man and yields started to fall where before they had risen steadily every year. There were all sorts of little problems, but one major one that they just couldn't fix: the heifers became very hard to get in calf. The Glasgow Vet School suggested that there might be a slight abnormality in the home bred stock's uteri which led to a stagnant pool of acidy stuff which made it hard for the heifers to conceive. Perhaps the old man should have bought in calving heifers and sold the home-bred ones. It wouldn't have cost that much if he had fattened the homebred heifers and bought in-calf replacements, and that way he wouldn't have been long in finding out.

But John Allan was past decisive action, and of course James Low couldn't act without him. They tried changing their bull but they wouldn't know if that was the source of the trouble until the new generation of heifers came to be bulled in about three years. They tried all sorts of mineral supplements but those didn't seem to make any difference. They stuck alkaline pessaries in the heifers to test the theory that it was acidity that was preventing conception. James Low was also past his best. He was only a few years off retirement age and looking forward to spending a bit more time on the bowling green.

Sadly, it wasn't just the cattle that were giving trouble. The staff who had looked after the place in the good times were dispersing. Bill Taylor, who had taken over as foreman when he came back from the war, had taken a place of his own up Deeside. Johnstone Riddle, the excellent foreman who followed Taylor, was left a bungalow in Ellon, but it was a condition of the bequest that he live in it, which would mean leaving Little Ardo. He decided wisely, took the house and a job working on the roads for the council. That meant his wonderful wife Ella, who had looked after the farmhouse when my mother was looking after her school in Aberdeen, was no longer available. Willie Adie, who had been second tractorman, got a job with a pension – as a postie.

When things were at their most frustrating the two old friends, the farmer and his grieve, met one afternoon in the close. After taking their problems through hand once more they decided between them that they would both retire. They would put the farm down to grass to let. Jimmy would stay on as Toonkeeper. He would count the visiting stock every day, put out cobs to them in the mornings when the grass grew thin in the backend, look after the fences and see that the steadings, which thirty-five years before, when he had first arrived at the place, he'd thought were already done for, didn't fall down altogether. But there would be no farming of Little Ardo on our family's behalf. It was a sad day, but I know of no evidence that my parents felt it. Jimmy Low seemed pleased enough, but Mrs Low felt it deeply. That was nothing to do with the family tradition on the place, she just hated the place being bereft of the stir in the close and in the buildings. For more than thirty years, when she had stepped out of her door she had been greeted by the curious stares of the beasts in the midden and a chorus of hopeful mooing from across the close.

So that was it. The cows were sold to Alan Buchan of New-machar for £107 a piece. I am glad I didn't see them go. For me it was a sad day. And it was sadder than I had imagined at the time. They were bought as replacements for Mr Buchan's herd, which had had a serious brucellosis breakdown. But he had bought too soon, and the Little Ardo cattle were dead within the year, slaughtered to prevent the further spread of the disease.

So the men left, the three cottages were let out and the Single Hoose, the one on the Big Brae, was kept for the farmer's son and his family so that they could come for holidays and visit without disturbing the peace of the farmhouse. Stephen Mackie of Balqu-hindachy, my mother's cousin, rented the potato quota and a field in which to grow the Little Ardo quota, and the rest of the farm was divided into two. Half was let to John Gyle of Esslemont for grazing his steers, and the other half was taken by Maitland Mackie Junior so that his dairy heifers could recover from wintering at Westertown. The farm now showed a modest profit, but showed it every year. There was no longer any worry attached to it or to the overdraft which had terrified John Allan all his time as a farmer, though in truth it was never more than £15 an acre. The next Allan to farm Little Ardo had far steadier nerves and he needed them, for his overdraft grew to a hundred times that much.

James Low and John R. Allan were now lifestyle farmers and the little farm on the hill stagnated. As the acreage necessary for economic farming rose inexorably, as the progressive farmers all round took the generous grants available and put up the modern sheds which were needed to allow the use of labour-saving equipment which was getting bigger and bigger, the old byre which James Low had described as rotten thirty-five years earlier and was hardly big enough to accommodate a wheelbarrow, was patched and the roans kept up and spouts kept open. But as there were no livestock in the winter, housing was not needed except for James Low's Morris Minor, John Allan's Ford Pilot and to keep their sticks and coal. The steading just sat there, a home for pigeons with no loons to hunt them any more.

Their neighbours just up the valley at Little Gight showed the two old men what might have been. The middling farm of Newseat of Ardo, 100 acres, some of which was so hilly as to be in use for nothing more than rough grazing, came on the market. It was John Mackie, the former grieve of Little Ardo who was now a junior minister in Harold Wilson's government, who had in 1965 introduced the Amalgamation Scheme. The idea was to encourage small farmers to give up the struggle and let their farms be amalgamated into economic units. The retiring farmer got some money to ease his retirement and the bigger farmer could get grants to help with any investments necessary to make the best of the new enlarged farm. . . . Roland Buchan bought Newseat for about £100 an acre and, because his inadequate buildings at Little Gight were even more inadequate for his extended empire, he got a big grant to put up a modern cattle shed, a high Dutch barn in the centre for a big silage pit and strawed accommodation for 120 beasts in a lean-to at either side. James Low had tried hard for many years to get John Allan to put a high roof over the Little Ardo pits, and here was Roland Buchan getting just that, up the road and in full view, and mostly at the taxpayers' expense.

But the new arrangement was ideal for the two old men.

John Allan had a steady income, free from the vagaries of crop failures, infertility, disease, and almost free of the overdraft which had worried him all his years as farmer at Little Ardo, while Jimmy still had the run of the place. He was able to continue to use his own time and the Little Ardo tractor to make a first-rate job of cutting the grass on the football pitch and develop his role as Village Character. At the back end, when the grass grew thin, James was able to add to his income with tips from John Gyle for putting out a bag or two of feeding cobs each morning to help to speed the stots off to market.

For Jimmy there was really only one blot on this idyllic scene. He didn't like the fact that the son and grandson of his hero, the Maitland Mackie who made him grieve before he was thirty, had taken half the Little Ardo grass. He really enjoyed young Maitland's annual visit when he came, describing him as 'a maist welcome visitor', on account of his custom of bringing a bottle of whisky to negotiate the rent for another year's grass. But he hated the fact that a rough place like Westertown, which he had sneered at for the previous forty years, was now in the position to take his carefully nurtured grass. He didn't exactly hate Westertown. I think he was jealous that he had only got the wee farm of Little Ardo to grieve, a mere 232 acres, whereas at Westertown there was a total of 700 acres. It might have been all right until John Allan came back from the war. At least till then he had sole charge and was responsible to the great Maitland Mackie, whereas at Westertown he would have had Mike Mackie to try to keep in order. But when the Hero returned in 1945 Jimmy was only a grieve at a small farm where the farmer might well have taken the job of grieve to himself. The fact that John Allan had no desire to and had not the experience to run the place without a good grieve helped a bit, but how James loved to criticise everything about Westertown. He thought the men there were ill-supervised, 'God Almichty,' he told me more than once, 'they yoke the plough and syne ging up to the head o' yon hill and sit doon be the dykeside and read the paper.' And how happy he was when a heifer which had gone missing at Westertown was found smothered under a soo of hay. It had eaten its way into the hay, which had collapsed onto it. James Low would have been a little bit sorry for the heifer, but he was more than delighted that Westertown had done anything so incompetent . . . and, better still, been found out.

143

And now Westertown was taking Jimmy's grass!

Still, James was now in an even stronger position to criticise the Westertown Mackies' every move. I was privileged to be in the close when the first batch of heifers arrived one spring to recover from a winter at Westertown on Little Ardo's sleepy braes. 'Here they go,' said Jimmy, as he let down the tail door of the float, 'the walking wounded.' There was evidence that Westertown had been short of straw, for the beasts were very dirty and covered with dangle-berries, but they seemed all right to me; I have always thought the wish to see animals clean was mainly based on the farmer's desires rather than those of the cows. Jimmy had taken great delight in phoning Westertown to complain that the grass at Little Ardo would be spoiled if they didn't hurry and send down their heifers to eat it up. 'The girse here'll seen be oot ower the dyke,' he had told them. And Jimmy was soon on the phone again to Westertown. 'You'll need to send Sandy Reid doon to worm that heifers,' he told young Maitland.

'Oh dear. Not looking as well as usual then.'

'Aye, jist aboot the same and they need worming.'

'Oh, yes. All right James. Let me see . . . Weekend coming up. I'll send Sandy down on Monday.'

'It's aa the same tae me,' said Jimmy Low, 'but if ye leave it till Monday Sandy winna need tae bring sae muckle stuff.' He had a fine line in exaggeration.

But he had no need to exaggerate the prowess of Westertown's Charolais bull. He was a magnificent, long muscular fellow who had been Junior Champion at the Paris Show when Mike Mackie bought him in 1962 for the first farmer importation of those great French cattle. Westertown Espoir was an enthusiast for his work but had no regard for fences: certainly the Westertown fences that were only guidelines for patient dairy cows, weren't nearly sufficient to keep him from the heifers. Jimmy was onto young Maitland again. 'Aye, Low, Little Ardo here. You'll need to come and tak hame fower o' this heifers. They're in calf.'

'Oh no James. That's not possible. They haven't been with a bull yet,' said young Maitland.

ABOVE. The Little Ardo loons, 1945. Back row: Joe Low, Billy Kelman and Jimmy Kelman. Front row: the author (note the fashionable underwear), Bertie Kelman and Albert Low.

LEFT. The author on his way to a great pay day (£4.0s.and 0d) at New Deer Show in 1954.

The author (far left) sneaking in amongst the 'Big Ones' at Dartington Hall School in 1954. Two years his seniors, they are: Diana Philcox, Marion Feld, Dickie Elmhirst, Julian Curry, John Worsley and Murray Cabot. (Nic Johnson)

The author (right) just fails to catch Franklin of Jedburgh in the 100 yards at Braemar in 1955.

The dance at the Beaton Hall in 1955. From the left: Tommy Sharp, Maggie Tawse, author, Alan Fraser, Betty Hay, Billy Ord and Elma Tawse. But why are both sexes at the same side of the hall?

At a University Ball with Mousie Clark in 1958.

The First XV of 1959 in front of a building they seldom visited, the Library at King's. Front row: George Masson, Tiny Duncan, Ox Nicol and Jonathan Moffett. Back row: John Raitt and John Munro. Middle row: Ian Phillip, John Boyd, John Stewart, Ray Hance, Brian Grassick, John Valentine (captain), Ian Macmillan, John Valentine, Donald Mackay, Neil Johnstone and the author. Back row: John Raitt and John Munro.

The author waits for the English class to convene with Ron McKay, Susan Macleod and Allan Macleod in 1958.

Fiona Vine didn't just have a flat, she had a wing of Drumduan, a dower house on Lower Deeside.

BELOW. The entire wedding party assemble in the gracious living room of Drumduan, Charlie and Fiona Allan's first home, in 1961. The author stands on the very spot on which he and Fiona took their vows in front of John R. Allan, Jean Allan, and Liz Vine.

TOP LEFT. The author holds Rosemary Anderson, daughter of Bill and Frances, who lived in the big house next door.

TOP RIGHT. Sarah, the author's first born and successor on the farm, sits on the top of Fiona's VW Beetle outside the cottage.

RIGHT. Ron Gazzard was sent by the *Dundee Courier* to take a picture of the young lecturer for 'Fiona', their gossip columnist. Like so many photographers Ron asked me to pick something up. The headline was 'Lecturer Supports His Family'. Oh dear!

ABOVE. When John Allan saw that his son had bought this house in Pollokshaws (it cost £4,100 in 1965) he said, 'You've fairly joined the bourgeoisie!'

LEFT. The author putts at Auchterarder in 1961, watched by Bert Maxwell, Jock Ritchie and Alex Bell.

LEFT. Ek Wallace throws the weight over the bar at Lochearnhead, c. 1965.

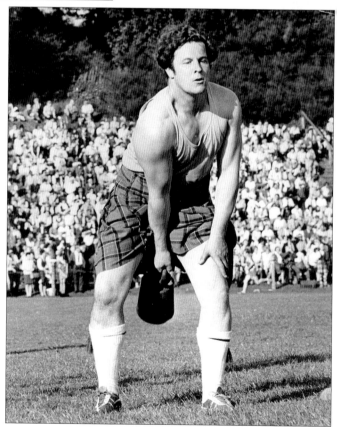

RIGHT. The author throws the weight over the bar at Pitlochry in 1965.

LEFT. Sandy Sutherland putts the shot at Blackford, c.1958.

BELOW. It's a Scottish Super-heavy-weight record! At least it was in 1970.

ABOVE. Although I had made it to the front row it seems I still wasn't good enough to get a dancer on my knee. The lucky ones are Arthur Rowe, left and Bill Anderson, right. Standing: Geordie Charles. John Freebairn, Alec MacEwen, Ek Wallace, Bob Aitken, Sandy Sutherland and Gordon Forbes, c. 1968.

RIGHT. The winning toss in the World Caber Championship in 1971. Keeping a watchful eye is judge Bob Shaw.

LEFT. That's the cup – 2 lbs of solid silver . . .

BELOW. . . . and that's the wages – Miss United Kingdom admiring the hundred-pound cheque.

ABOVE. The author with the caber at Dundee in 1971 watched by Gordon Forbes, Bill Anderson, Bertie Paton, George Donaldson, John Freebairn, 'Hurricane' Ed Weighton, Bob Aitken and the great commentator, Jimmy Miller.

RIGHT. George Halley the Blackford blacksmith, great games organiser and judge, gives the author's spike a friendly 'tramp-in' before his winning throw of the heavy hammer at Crieff in 1972.

The author addresses the caber at Braemar in 1968.

The winning throw at Aboyne in 1969.

ABOVE. There aren't many laughs at the Games but I used to get one if I jumped over the high jump without removing my kilt. I even claimed a kilted high-jump record of five foot nine at Arisaig in 1967.

RIGHT. George Clark putts the stone at Aboyne about 1945.

LEFT. The young lecturer brings the house down by throwing the weight over the bar with his final throw, Kelvin Hall 1962.

BELOW. The McVities team arrive at Tokyo airport. They are: Margaret Murphy, the author, Dougie Duncan, Dave Prowse, David Webster, 'Wild' Ian Campbell, Billy Forsyth (then world Highland Dancing Champion), Clayton Thomson and Arleen Stewart, who all the Japanese announcers called 'Andy Stewart'.

The author with the cleaners at Tokyo's Toshiemein Park.

The McVities team visit a professional sumo club where training had been delayed by three hours so we could still see some of it when we got there at 9 a.m. The team from the left are; bus driver, Bill Anderson, Billy Forsyth, Ian Blair, Margaret Murphy, interpreter, 'The Highland Herculese' Graham Brown, Dougie Duncan, Charlie Simpson, the author, Arleen Stewart, Dave Prowse, 'Wild' Ian Campbell, David Webster, and Mr Hurry-Hurry-Bus-Leave-Half-Past-Eight.

ABOVE. The Australian caber before Arthur Rowe broke it. It is supported by David Webster, Gordon Forbes, Colin Mathieson, Arthur Rowe, Bill Anderson, John Freebairn and the author.

LEFT. The author with Mr Hurry-Hurry-Bus-Leave-Half Past Eight.

BELOW. The first four Simmentals to arrive in Aberdeenshire. From left to right they are Sophia, Heidi, Gertrude and Gretel.

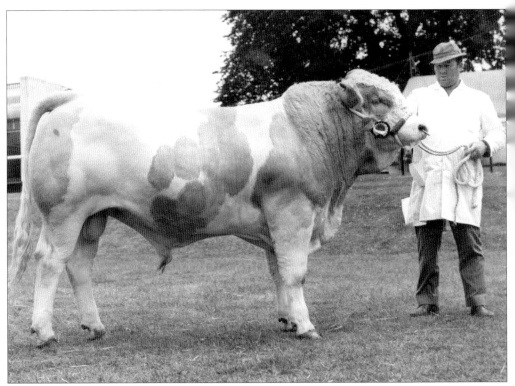

Ardo Dante, Champion at the big Simmental bull sale which was at first held at Edinburgh. The stockman as he took the Junior Bull award at the Royal Highland Show is the famous show cattleman and Glenprosen bon viveur, Bertie Paton.

With Gordon Paterson after grooming Ardo Glendevon in 1978.

Ardo Figaro, Perth supreme champion and record breaker for weight gain, with Gordon Paterson.

Some of the exotic menagerie at Little Ardo in the late seventies being entertained by a guy trying to sell tapes of bothy ballads. They are Simmental, Gelbvieh and Marchigiana.

The first complete dispersal sale of the Ardo Simmentals. The author is flanked by auctioneers Roley Fraser and John Thornborrow. Jack Clark is standing is standing on the right and Simon Fraser is on the left.

Mrs Mary Dingwall-Fordyce presents. Methlick Football Club, bottom of the Buchan league, beat New Pitsligo, top of the league, on penalties after a two-all draw, and lifted the trophy under the floodlights at Longside in 1983. The heroes are – back row: Mikey Taylor, Dod Barrack, Doug Blair, Whitey Bonner, Jimmy Duncan, Kenny French, Bill Bonner. Front row: the author, Andy Bonner, Hamish French (later of Dundee United), Allan Low (grandson of the great Jimmy Low, grieve at Little Ardo) and Brian Gray.

Judging the supreme Boran Championship at the Nairobi International Show. Well, I was in Africa anyway.

Camping at Lake Naivasha with visitors from Scotland: Mary Levie, Susie Allan, Karin Mackie, Fiona Allan, Ann Wright, Julie Eagles, the author, Allan Wright and Jimmy French on the way to climb Killimajaro – 19,000 feet without oxygen.

The author and Maitland Mackie get set for another day hiking in the Masai Mara while Kirstin Mackie tests the water of the Mara River. Photograph by Dr Halldis Mackie.

The author with one of the Jersey cows who were part of another scheme to gang agley, 1990.

Among newly weaned piglets – an odd place to lie down though the straw was clean, c. 1994.

The winners of six Bothy Folk Song competitions meet for the overall championship at Elgin Town Hall c. 1995. They are from the left: Gordon Easton, Joe Aitken, the late Tam Reid, the author, Scott Gardiner and the late Ian Middleton.

Raising money for Barnardo's. Beechgrove gardener Jim McCall, the author, Aberdeen International centre-half Brian Irvine, Frank Gilfeather. Sitting (Sir) Alex Ferguson and Kitty Pawson.

At a joint book signing in John Smith's in Glasgow where *Farmer's Diary* volume 3 outsold Edward Heath's memoirs.

The author and Fiona at the retirement housie they built in the neuk of the Hill Park.
(© Jim Henderson)

Sarah, the sixth generation of us on the hill, and her husband Neil Purdie are well settled in the farmhouse at Little Ardo. (© Jim Henderson)

'Well that's bloody funny, for they look in calf tae me and there's twa feet stickin oot the back o een o them.' I cannot always be sure of my stories, but I can be fairly sure of that one, for both parties told me it and both enjoyed it hugely. That is one thing that may be said about the Mackies on the credit side of the ledger: they do enjoy a story against themselves – even when it's true.

Later that summer we were holidaying in the Single Hoose on the Brae and in the night we heard a great roaring down in the Howe. It was one of Westertown's dairy heifers. The four had all been taken back to Westertown but this was a fifth that even Jimmy hadn't spotted. If it had been anyone else I would have expected him to keep quiet about it and to sell that calf to best advantage at Maud, but James Low would not have. Still, I don't know what did happen to that calf.

Anyway, the arrangement with Mains of Esslemont taking the best arable bits of Little Ardo for his fattening steers and Westertown taking the braes for their dairy heifers lasted until 1973. That was when I decided to give up the university life and return to the plough. I might have done so sooner had the farm been bigger or had I not been afraid of tying myself into a debt-ridden and profitless farm that my father had only sustained by his writing. But in the early '70s my parents still needed the house and an income from the farm. It was all right for me to think that I should be at home seeing to the modernisation of the place, but I could see no way the farm could match a senior lecturer's pay, plus Fiona's profit from her pregnancy-testing laboratory (about which more later), at the same time as finding the investment necessary for new buildings and machinery and maintaining my parents' income.

CHAPTER THIRTEEN

Exotic Cattle

1969–76

Then came Gelbviehs and Limousins and Blondie Aquitaines,
Normandaise and Murray Greys and Anjou Mains.
There were MRIs and WRIs and Marki power,
Roamin-in-the-gloamin, Rotguts and Pinzgauer.

It seemed that Jimmy Low's fine retirement jobbie overseeing the farm's slow decline and his discomfort with being second fiddle to Westertown would continue for long enough. But in 1969 something happened that was to change the whole equation. I visited my cousin James Mackie at the Bent of Laurencekirk and saw his Charolais cattle. He told me they were worth £3,000 apiece. Now, my academic economics was nothing to do with practical decisions like how to make a fortune in farming, it was to do with wealth creation and distribution in a national and inter-national sense. But even I, who was used to thinking in terms of the greatest good of the greatest number rather than mere balance sheets, could see that there was the possibility of profit here. If I bought a shorthorn cross Ayrshire cow for £200 and she produced an Aberdeen Angus cross calf I could hope to sell it fat at Maud for £200. If the cost of feeding it and its mother were £190 that would be a profit of £10. But if I were to buy a Charolais cow for £3,000 and she had a £3,000 calf, how much better would that be? They might eat twice as much, but that would still allow a profit of £2,600 . . . every year! Times as many cows as you like!

This was clearly what the small farm of Little Ardo had been waiting for. With cows like that it could generate enough income to keep John Allan in the style to which he was accustomed, while rebuilding the steading and giving his son the standard of living to which he aspired.

I wasn't so naïve as to think my plan was foolproof or to go hell-for-leather for it, but I did join the Charolais Breed Society and signed up for the next importation from France. By early 1970 four Charolais heifers were rattling around in the old steading. To help finance the deal (as they were to cost £2,000 each) they were bought in partnership with Neil Massie, whose Blelack herd were among the pioneers of the breed.

Joining the Charolais Society was my first step in farming but oddly the Charolais were not the first home. I managed to get some of the very first importation of Simmental cattle from Germany. And getting on that list was perhaps the best bit of farming I ever did.

I had been up Glenprosen visiting Bertie Paton's father-in-law, Archie Whyte, whose Spott herd of Blacks were among the founders of the herd book and had won two gold medals at the turn of the nineteenth century. Archie was amused by my intention to import Charolais and he showed me the *Dundee Courier* report of more exotic cattle coming in from Europe: 'There's more foreign cattle coming in – "Sementimentals". Aye, they're coming into Dundee.' The milk boards wanted them to improve the meat production of the dairies by putting more beef onto the calves of the less good milkers, whose progeny were not required for replacements.

The Scottish Milk Marketing Board (SMMB), who wanted ten bulls for their great artificial insemination (AI) station at Perth, under the chairmanship of Skerrington Mains himself – the great and terrible Willie (later Sir William) Young, were organising the importation.

Not having any idea how these things worked, I phoned the SMMB headquarters in Paisley and asked if I could get on the list for some of these 'Seminal' cattle that were being imported from

Germany. The voice at the Paisley end of the phone roared with laughter. 'Oh, dear me no, Mr–eh–Allan. The people on the list have been waiting for years and the cattle are well oversubscribed. In fact, the successful applicants are having a meeting tomorrow to finalise the project.' I asked if the meeting would be at Paisley and was told it was set for ten o'clock the very next day. So I thanked my informant, who apologised for his inability to help me. I couldn't see how he could have been more helpful. I decided to gatecrash the meeting, but to crash the gate so quietly that no one would realise I was there.

I was very excited, especially after I had phoned James Mackie, who was one of the successful importers who had made a lot of money out of Charolais, to ask if there might be a market for the progeny of these 'Sentimental' cattle. Indeed there was. In fact, Danny Thomas, to whom James had been selling Charolais for his Four T ranch in Texas, would give me £2,500 each for the first calves off the importation . . . but of course I had no hope of getting on the list.

I arrived just in time for the meeting. I didn't want to be standing around giving everybody a chance to ask who the hell I might be. And so many of the giants of Scottish livestock breeding were there. Their names were on a list of importers of Simmental cattle which was circulated. There was George Anderson, son Alec Anderson of the shorthorn breeder from Burnton; Jim Swanney, who made a fortune for the Milk Board out of several breeds of European cattle; Frank Young, the Galloway breeder, was there looking for some new blood for the Congeith herd; Tom Barr, the Ayrshire dairyman and Blue-faced Leicester breeder who was to produce the first two winners of the supreme award at the big Simmental bull sales which at first were held, not in Perth, but in Edinburgh; Gerry Rankin, international shorthorn judge and director of the Boots pension fund cattle and Robin Forrest who, with his brother Barclay, seemed to farm half the Borders. In the chair was the great Willie Young himself. I expect he was looking avuncular, but to the interloper he was a fearsome sight.

After some explanation about how the purchase and importation of the cattle would be handled, the chairman said without any apparent malice, 'Now is there anybody here whose name is not on the list?' My heart sank. I said that I had not appeared and that I was Charles Allan of Little Ardo in Aberdeenshire. That was OK. The secretary took my details and I appeared to be in. 'Now,' said the chairman, 'some people here haven't said how many cattle they want.' Most people said six, so I said six when it came to the turn of 'Charles Allan, Little Ardo'. I rather liked the sound of that. It was the first time I had heard my name announced as though I were already standing in the shoes of the Yulls, Mackies and Allans who had cared for the family farm since 1837. The only other thing I learned there was that the heifers were to cost an average of £400 each, perhaps twice what you would have paid for a good cross-bred heifer to make into a beef cow on the hill. Having already paid out £2,000 apiece for Charolais heifers, that seemed cheap to me though I knew nothing much about it – but there was one who thought £400 an outrageous price.

My pride at the cattle venture must have rubbed off on our four children, who thought they had surely got one over on their pals at Robslee Primary School beside Rouken Glen Park. But they were sadly disappointed. When they announced that their father was importing Simmental cattle they were dismayed to be told that Stephen Lipp's parents already had Simmentals and so did the parents of half the other children at Robslee. I hope their fathers had had a better reception when they went to see the family bankers about the financial arrangements for acquiring the German cattle than I got at Methlick. My life being in Glasgow with my university and my family, I kept my overdraft with the Bank of Scotland and had lost touch with the family bank in the village. The Yulls had kept such money as they ever had with the North of Scotland Bank, which had now become the Clydesdale and North of Scotland Bank. The new enterprise was to be called Ardo Pedigree Cattle, which I thought very grand.

I explained to Mr Clark, the manager, that these cows weren't in demand just because they were good cattle, although they were the most numerous purebred cattle in the world after Herefords. The

Americans were desperate to get their hands on Simmentals from Britain because foot and mouth disease was endemic all over Europe, so they could not import Simmentals into the States in case they brought the disease with them. When the British government licensed importation into Britain the Americans still couldn't get them into America, but they were allowed to import their progeny. If the British were to take the chance and import the parents, the Americans would be safe enough to take their calves. 'Oh yes, Charlie. I see. Very good, that is just the sort of business the Bank would like to be involved in. Now, six heifers. How much will you need to buy them?' When I told Mr Clark that they would be £400 each he was flabbergasted.

'Fower hunner poun for a heifer! Oh no, Charlie, the bank couldna hae nithing tae dae wi that.'

In the end, because of failures to pass tests in quarantine we only got four heifers, but the Bank of Scotland in Glasgow just added £1,600 to my already thriving overdraft, and in less than two years they had collected four cheques for £2,500 from Danny Thomas. The next generation of calves did better still. The Australians and New Zealanders also wanted Simmentals but had to wait for the British calves to be available. One of those second calves was bought by a Swede for export to New Zealand for 10,000 guineas – £10,500.

The Clydesdale and North of Scotland Bank maybe didn't think that was the sort of risk with which they should be associated, but to me it looked like the answer to Little Ardo's prayers. With that sort of money coming in, a keen farmer would soon have the old steading refurbished, and the place restocked and equipped with the sort of machinery that would attract and keep good men. If the extra 150 acres the little farm on the hill so desperately needed came up, none of the other neighbours would stand a chance.

It was an exciting time. Old James Low, who was still caretaking for John Allan, had already said he would gladly look after four Charolais heifers. And now there were to be four Simmentals. Eight cattle were nothing to a man of James Low's expertise, even if he was sixty-six. But then there was another Simmental heifer, and then four more. Not long after that, four Gelbvieh cattle arrived from Germany. And then to speed up the process and the cashflow we sent one of those for transplant.

Eggs were taken from the expensive pedigree heifer and implanted into cheap cross heifers to give more high-value calves per high-value mother. The deal was the transplanters didn't charge me anything but they kept half the successful transplants gratis. So Jimmy had another four Gelbvieh calves to deliver. I sent them a Chianina on the same deal.

When the first of the Simmental heifers calved, with James expertly hauling it out with the wire stretching pullies he had used for years to give one man the strength of eight, he had something special on his hands. Already sold to an American who would take it to New Zealand, Ardo Antipodes was the first Simmental calf to be born in Aberdeenshire.

And Jimmy was still holding the fort alone when our first heifer available for export came on the market. By this time I had wasted a couple of thousand pounds buying from their home county four South Devon cows, the biggest British breed and so thought best for breeding with the Simmentals. Anyway, one of those had timber tongue, a very nasty and often fatal disease which, among other things, makes the beast's tongue go as hard as wood. The Swede who came to buy the calf was very taken with the way the old man was tending to his stricken cow, covering her with blankets and straw and feeding her whisky, so that when his wife said, 'We should also take the cow,' he replied. 'I think we should take the cattleman as well.' Jimmy liked that, though he knew it was just charm.

Old Maitland Mackie was eighty-six in 1971 when we had our first Simmental calf. It wasn't just our first. I was proud of being first in the county, but that was nothing to the pride of showing my grandfather, who had seen so much, something he had not seen before. He had been such a pioneer and just loved to show people his own successes, and even his failures. Though he was getting very old he was erect, had no limp, and saw the humour in the earnest young man showing him a ferlie.

Maitland Mackie had reached the stage when he couldn't hold a long or complicated conversation, but he still knew his cattle and he still knew how to tease his grandson. The old man's sense of humour was still intact. It was with great pride that I took my grandfather to see my pioneering calf. He and his mother were in the old bull's pen. In this dark dungeon of a place, which the old man himself had built fifty years earlier, I explained all about it: this Simmental was the first to be born in Aberdeenshire, and I had a buyer in America who was anxious to pay me £2,500 for him. The old man made a good show of being impressed by the figure though I don't know that he would have believed it. He said it was 'a nice calfie', and then with a wonderful twinkle in his old eye said, 'But it's a bittie high in the tail isn't it.' Indeed it was, but, as I explained, these big European breeds are better to have the tail raised so that it doesn't interfere with the birth of their huge calves. Old Maitland then moved round the pen a couple of steps and, looking again at my calf said, 'Now, here's a nice calfie,' and then, eyes twinkling in just the same way, 'but it's a bittie high in the tail, isn't it?' I explained again that with these new breeds having a smooth straight top line was less important because it made calving easier if they were a bittie high in the tail. He listened carefully, nodding at all this wisdom coming from his grandson, who was very clever at the university but who knew so little. Then he moved again a couple of half steps round the pen. 'Now here's a nice calfie,' he said, 'but he's a bittie high in the tail . . .'

It was a very affecting occasion. I felt very close to the old man. I was very glad that I had known him so well when we only had to say something once and when he expected good answers to his questions.

That calf was also a great source of pride to James Low. He had never been at a farm known for pedigree cattle, and to have one sold for £2,500, and the first of its kind to be born in Aberdeenshire, was an excuse, not that he ever needed one, for a good deal of heavying. The venue for most of that was outside French's, the general merchants in the village, to which he went every morning to collect his paper. A natural leader and a grieve for three-quarters of his working life, Jimmy was now the grieve of the collection of old men who paused to have a news as they collected their papers at the back of eight each morning. The odd busy farmer would stop and add his pennyworth about the mess the government was making or what had gone wrong with Bertie Taylor's organic oats. With this 'Sentimentiment . . . this foreign calf' James had gathered quite a crowd who were doing their best not to be impressed by James Low's achievement. And it was his achievement, for he was the only staff and the only management at Little Ardo – except for the lad in Glasgow who phoned up every day about dinner time and asked daft questions.

Among the crowd was our neighbour, James Catto, who farmed the much bigger Auchencrieve across the Methlick to New Deer road. Whereas Jimmy Low had a dozen cows, and let out the rest of the farm for grazing, Auchnies had a big double dairy byre, twice the acreage of arable and a priceless resource in the Belmuir, 1,000 acres of flat heather moor complete with hill-cow subsidy, the biggest of its kind in Aberdeenshire.

In patronising tones James Catto asked about 'yer calfie' and then said, 'I'll gie ye a fiver for it.' This got a good laugh from the old men who knew of Auchnie's reputation for never spending a fiver unless for a bargain, and they had all felt the weight of Low's tongue and his personality.

But Low was fit for that. 'I micht gie ye a *look* for fiver,' he said, 'if I thocht ye had een.'

Soon these Charolais and the Simmentals were coming, if not exactly thick and fast, at least fast enough to show handsome returns to the telephone farmer in Glasgow and his cattleman James Low in Methlick, who was on 10 per cent. Those two breeds did have something extra to offer the British herd. The Europeans had bred their cattle to pull carts and ploughs for far longer than the British, whose agriculture had been far quicker to specialise. We used horses for draft and kept the cows for beef, leather and milk. Now, the specialisation in farming had, just as it had in industry, been a great stimulator of change, but there was a feeling that there had been a cost. Our beef cows had became too small and too fatty, so these big rough draft cattle from Europe were wanted to give us more lean muscle. That was a clear advantage for the Charolais. Then the Aberdeen Angus, Herefords, Galloways, Highlanders and even the Shorthorns had lost some of the milkiness of the cows, and that was where the Simmentals could add something. All over Europe this big brown cow with a white face was used as a dual-purpose animal. When it calved it went straight into the dairy to be milked twice a day. All the females were milked for one lactation and the least milky 90 per cent were then slaughtered for beef. They were not regarded as farrow cows as in Britain, but as prime beef, qualifying for all subsidies and sold at the full price. So the Simmental could offer almost as much beefiness but also the prospect of breeding some milk back into the British hill cows.

The influx of new blood from Europe was by no means finished in 1970, when the Simmentals arrived. The Americans were willing to pay big money to get, it seemed, anything bovine from Europe, and though we didn't know it when I was a boy, the Europeans don't go in for cross-breeding and there are a large number of pure cattle breeds on the continent. The most unlike anything we had here was the Chianina, whose home is in Italy. It is the tallest bovine in the world, so the Americans must have them. The first men from the North-east to see the Chianina were a posse of farmers who went in 1970 to the Paris Show. Their reports astonished us. Alec Stott and Andy Willox, who were always at everything if it was in Aberdeenshire, had somehow appeared in Paris. Now Andy was a little cheery round man and Alec was by comparison tall and only a little less round. They described this Chianina bull as being so tall it could 'shite ower Stott's bonnet' while Andy could 'traivel throw atween its legs without booin doon'.

So naturally the lecturer in Economics at the University of Strathclyde put his name down for these monsters too, as he had done for every breed in Europe. He was even on the list for Magyar Zurke cattle from Hungary, which at that time was still behind the Iron Curtain, so they never made it to Britain. At any rate, the day came when I had to tell Jimmy that he was to get another six cattle: four Gelbvieh (golden cattle) from Germany and two Chianina from Italy. I could see that Jimmy was reaching full capacity after two years of an expanding herd of liquorice allsorts, all to be AI'd and with different breeds of bulls. 'Fit wid happen if I got nae weel?' he asked with some indignation and just a hint of desperation. That was a good point, for he was alone on the farm. It was clear that his fine retirement jobbie was not living up to expectations.

The Chianinas, as well as being the tallest cattle in the world, were, if you believed the publicity, the original draft animals used for ceremonial jobs, including drawing the carts in which the Christians were hauled into the arenas of ancient Rome to be devoured by lions. The story went on that they have been bred pure ever since in a small area of Italy, so that when they are crossed with other breeds their calves show hybrid vigour second to none. I don't really remember how it happened, but I landed with one and a half Chianina heifers, the extra half belonging to Bertie Paton, by now Archie Whyte of the Spott's son-in-law. A Chianina split between us didn't seem to make much sense and, as these brutes were selling well, snares were set to get an American to take one off our hands. Of course, Americans couldn't take these cattle to the States, but some were buying them and getting others to breed them in Britain and then the progeny were exported. Because Jimmy Low was so sceptical about the Chianinas, Stephen Mackie at Balquhindachy, just three miles up the road, was looking after my heifer and a half. They had cost £2,700 each and Stephen was astonished when the Strathclyde lecturer told him to ask £10,000 for the lesser of these two skinny creatures. Long experience in the cattle trade made Stephen decide to give himself a bit of room to manoeuvre and, hiding the better one, he asked the first ten gallon hat for £12,500 for one heifer. The American agreed so readily that Stephen was sure he had not asked enough. I had a very good job in Glasgow and yet this one heifer was worth two and a half years of my pay as a senior lecturer.

James Low had been scandalised when I described these huge cattle. He wasn't impressed that calves got by a Chianina bull on native cows would produce a cross of extraordinary vigour, would grow like nothing we'd ever seen before and, having thin bones and a reluctance to put on fat, the carcases would yield a high percentage of lean meat. But Jimmy Low was having none of all that. It wasn't the Christians he was worried about, it was the Scottish cattlemen who would have to manage these huge brutes.

'But, Jimmy' I said, 'I've seen them at a cattle show in Italy and they're as docile as you like.'

'Jist wait or the buggers calve,' he said, with that look of perfect disgust which was a bit of a trademark.

I sold the first of the Chianina cross at Maud. Stephan Mackie had put some of the semen I had bought on some of his less good dairy heifers and I would get the calves, the bulls for fattening and the heifers for grading up to pure. At any rate, even I could see that the Chianina market was to be for heroes only and so I fattened this heifer off. It did create a stir. It must have been a foot further from the ground than any heifer of a similar weight. A rangy black brute with a rather sweet face, I thought. It was bought by the late John McIntosh, the butcher with whom I used to share a joke on a regular basis at the fat ring and who liked to try things that were different. When I saw him the next Wednesday at Maud I asked John how the Chianina had killed out, very professional-like. 'Jist like a fuckin cricket bat,' he said without a hint of a smile. I didn't press him further.

Despite that unpromising commercial experience, the ease with which the pure Chianina sold, and Jimmy Low's desperation for the flow of cattle to stop before it buried him, convinced me that I could now go home to the plough. I had already, at Jimmy's suggestion, managed to fee the excellent Willie Adie, who had been such a mainstay in John Allan's great days on the farm, to come back again as cattleman. Of course, there could never be a question of Willie being as good as James Low, but the old devil was desperate for help, or at least the odd day off. I did think that while the money the exotics were throwing at me gave me a chance to modernise the old place, I would really have to be there to see it done.

It wasn't that I wasn't enjoying the university life, but I had come near to a crossroads. I enjoyed the teaching very much, and the contact with the students. I enjoyed my bits of spying for the government and my research leading to publication in academic books and in journals, but I had reached as far as I could without becoming a professor, and a professor I didn't want to be. Oh, I would have liked the prestige, and Fiona would have been pleased to be a professor's wife, but in those days professors hardly did anything except sit on committees. There were committees to supervise the development of courses and to put up suggestions about how the research of the staff should proceed. There were interviews to conduct for new staff and more meetings to appoint external examiners for the students' degree examinations; and professors were on the university court which ran the university. Professor Alexander, my boss at Strathclyde, was busier in industry, trying to save the Clyde Shipbuilders, just as Professor Campbell, at St Andrews, had been more involved in trying to save the jute industry than he was in teaching and he did no research. Both took the first-year students for lectures, but those were classes of up to 600. There was hardly any contact with students in manageable numbers. Although my career was going well and although I was trying my damnedest to get on, I knew that if I succeeded with the step to a chair it would not suit me.

Then again, there was a philosophical reason why my study of Economics was less satisfying than it might have been. I began to see that my Economics, while perhaps good for the brain and even for the soul was, for the working of the world, pretty well useless. What I spent a lot of my time doing was trying to fix on a framework for decision-taking that would enable us to make good decisions. With the aid mainly of geometry and a little algebra I tried to show how we could decide if one set of circumstances was more desirable than another. Would a change be a good thing if it made one person better off and no one worse off? Well, yes, but what if the change made one person better off and two people worse off? Well that would depend on how much better or worse the changes made people. A change that made ten people a little better off but made one person absolutely miserable wouldn't necessarily be so good. And what about a change that left no one worse off and nearly everyone better off but the few who had not benefited were so furiously jealous as to become suicidal . . . or even homicidal? Well, you can see that a young man who had gone into Economics to help the world understand itself and to make sensible decisions about how to order things might be upset at finding that he couldn't even decide how we should decide. You may agree that a man who at thirty-three is reduced to worrying about such nonsense had better get back to the farm as soon as possible.

There were other things. The children were happy in Glasgow but they weren't as streetwise as their fellows. They were daft about football but they foolishly let it be known that they supported Aberdeen. That was not a good idea and left them open to abuse from Rangers and from Celtic supporters. They had to pay protection to both sides to allow them free access along Hillside Road to the shops on Saturday mornings to spend their pocket money. That contrasted unfavourably with the way every-one stopped them on their way to shop in Methlick and chatted with them. Then the athlete was wearing past his best and thought hard physical work on the farm might revive his Highland Games career. It would have to be better for a declining athlete than being stuck behind a desk in Glasgow deciding how to decide.

But the main reason for giving up the university was the desire to do something to restore Little Ardo to something like the past glory when William Yull built the Old Barn and all the neighbours came round to see it because they had heard that the built-in threshing-mill had been installed back to front so that all the straw and grain landed outside instead of in the barn; when John Yull needed three gallons of whisky to put in the New Year, and when Maitland Mackie put up his new dairy byre and covered the braes with White Wyandottes. With a dozen cows producing a calf worth thousands most years, and as many more on their way from Europe and a seemingly endless supply of gullible Americans, we would be able to live the life of gentlemen farmers while doing the restocking, rebuilding and re-equipping for which John Allan had not been fit.

It was exciting – even if it was an illusion.

CHAPTER FOURTEEN

Return to the Plough

1974–77

Of aa the airts the wind can blaw
It canna blaw like me.
I've got my chance I'll show them noo
What a richt fairmer can dae.
The tenants got their orders,
I wadna be sae daft.
I went to see the banker and got me an overdraft.

John Allan had been a good husbandman. He was generous with the muck and the lime and never put on the nitrogen when a compound could be found. He made sure that the dykes and the steadings were kept up, the weeds and the rats were kept down, the drains were kept open and the overdraft was kept down. My father didn't do as his brothers- and sisters-in-law had done. He hadn't sold the farm and leased it back to raise money for high farming. He didn't solve the long-term problems about the size of the place in relation to the economic needs of modern agriculture, but he did make one enduring contribution that transformed the place for the better. When he came back in 1945 from seeing to Mr Hitler he found one of the barest farms in Aberdeenshire. The thirteen mature trees and the scraggy hedge round the house and garden were scant protection against the wind that came at us from all directions unobstructed. As soon as he was home John Allan, with what he was kind enough to call 'help' from his six-year-old son, set about creating a beautiful setting for the eighteenth-century house among a delightful variety of trees. There was a lot of scoffing at his choice of poplar trees to line the Big Brae. 'Funcy stuff like that winna grow at Little Ardo.' Sixty-five years later you can concede that the sceptics may have been half right, for, though they grew very quickly, straight and tall for fifty years, they are already beginning to succumb to the west wind. The willows were more successful and we got a wonderful surprise in 1990. Four years after he died, we took a crop of cherry plums off the trees John Allan had planted among the willows. When he became farmer here, birdie's pizz and the odd rasp was the best you could glean on the Big Brae, but now we enjoy his bounty in plums every year.

For all my academic success, for all that French, Spanish and Japanese students could read my book on taxation in their own languages, I was elated by the thought of gathering on the braes the stones my great-grandfather had missed, of cementing the floor of the barn my great-grandfather had built, of modernising the piggery my grandfather had built and adding profitability to the fertility my father had fostered in Little Ardo's fields.

You would not need a degree in mathematics, economics or even accountancy, or even much common sense, to wonder that I didn't cash in more of those cattle and those huge prices. When we got £12,500 for one really awful Chianina heifer, it would surely have made sense to sell the other one which was only pretty bad? Well, yes, I should have. Four years later I was selling my (by this time) seven Chianina females for £2,750 for the lot – little better than commercial prices. I knew quite well the exotic boom wouldn't last long but I did think that the big prices would last a good deal longer than they did, and, if I had sold more, I would have been hit by a colossal tax bill. At worst the government would have needed 87.5 per cent of the profit had I sold that second Chianina. I didn't speak to him myself, but Maitland Mackie, my cousin, told me that Al Philip, a North-east exile making his fortune in cattle in the United States, was interested in giving me £20,000 for those first four Simmental heifers the Methlick banker had been unwilling to finance at £1,600. But the exchequer might have needed all but £2,500 of that. Far better, I thought, to keep my seedstock and live royally off the sale of a few calves every year. That was what we decided to do, and with that intention I gave in my notice to the University of Strathclyde.

Of course, I could have avoided a good deal of that tax by investing the money quickly, 'rolling it over' as we used to say, into all those improvements I hoped to make, or I could have sold out 'on the herd basis', which could have allowed me to get together a great herd of good commercial cattle or even bought a new dairy and restored the place to its postwar heyday. But I couldn't do that over the telephone. That sort of radical surgery would have to wait until the new farmer was in place. By the time I had served my notice in Glasgow, got my clock, sold the house and moved, the whole daft exotic market had blown up in my face. The oil crisis of 1973 brought the American interest to a stop all at once. Suddenly all these strange cattle were going to have to find markets in Britain or perish. By the time I returned to the Little Ardo the huge value of all those beasts was gone. The issue of rolling over great profits disappeared. I was clearly going to have to be a good farmer like my ancestors and I wasn't by any means sure I knew how to be that.

So I answered the call of my ancestors to the Little Farm on the hill. That was all very well, but what about Fiona, the wife who had given up her academic career at the University of Aberdeen to support mine, wherever it took us? What did she think about being uprooted from the life she had made in Glasgow? And her husband giving up the ladder up which she had helped him to move smoothly towards a professor's house in the Chanonry in Old Aberdeen, in Kelvinside or the New Town in Edinburgh? And she had made a very good living for herself in Glasgow. Her husband had not made enough for comfort with four children growing up and needing everything, so we had managed without. As soon as she was recovered from her fourth baby she got a part-time job as a chemist at Mearnskirk Hospital. It was difficult to fit in the travelling, the school runs and all the business of running a big house with six people in it. So Fiona set up a business of her own, at home, and a right good business it was. One of her jobs at Mearnskirk had been to test urine samples for pregnancy. It was a simple test, for which her chemistry/maths degree and hospital experience made her well qualified, and she saw that there might be a demand for a private service. Her friend Dr Andrew Thomson, a local GP, encouraged her that there would indeed be a demand. The standard procedure was to take a sample to your doctor; he would send it to the hospital where it would be tested, and less than a week later you could go to the doctor again for your result. She thought some would prefer to visit her at home, pay £2 and get their results in three minutes.

Well, she was quite right and soon the chemist, housekeeper, wife and mother was earning more than the husband with his fancy job at the university. The poor woman had to leave all that behind when we left Glasgow. While someone would need a test in Methlick every now and again, there were neither the numbers nor the anonymity that had been important to her business success in Glasgow. It was understandable that I was dragged home by the peasant's instinct for his land, but it wasn't Fiona's ancestral home, so why should she give up so much once again to come with me?

Well, it is true that while the farm wasn't her heritage it was the heritage of her children and she was just a very selfless wife and mother. She again left her good career to follow the one I had chosen, this time on a downward trajectory. I don't know that I have shown my gratitude sufficiently.

One thing though, very conscious that this move was my doing, and had many arguments against it, and knowing that there would be difficult times: I made a little promise to myself and unspoken to Fiona. I would never come in from a hard day's work and inflict a blow-by-blow account of all my misfortunes on my wife. I wouldn't throw myself into the easy chair with, 'What a day I've had!' If the stots had got into the barley and the pigs had got out of their weaning shed, if the tractor had stuck in the moss and had taken me half the day to recover, if the banker wanted to see me and the bank rate had been increased, I would not inflict that news on her. It was partly that I thought it would be unfair, because though they affected us both, those troubles were of my making, and partly because I believe, contrary to the old saying, that a trouble shared is a trouble doubled. It may be a great thing to tell your troubles to a priest and be forgiven, but within marriage I think the best thing is often to share any good news that is going and let the disasters seep out. Happiness with another tends to feed on itself and grow. After a while of savouring the good news, and having a laugh, the bad news often seems to fade away.

You will have gathered that while I was the partner with the fancy job in Economics, Fiona was the one with mother wit. She took a fairly brief look at the situation I had landed my family in at Little Ardo and decided she would need to step in again. She had given up her teachers' training course when she had started as a lecturer at Aberdeen University fourteen years earlier, so she registered to finish her training in Dundee and was soon teaching maths at Turriff Academy. While I struggled to cover costs on the farm she again became the breadwinner of the family. She was very good at it too, and not without ambition. After five years at Turriff she got a grant to retrain in computers, came first equal in Scotland in the top further education exams for postgraduates and was asked to stay on at the College of Commerce as a lecturer. So she was able to win quite a bit of bread.

One who had the greatest difficulty in understanding my move into farming was James Low, now on his forty-third year of service to our family, mostly as grieve, but also as guide, philosopher and friend. He had brought up his five children with the primary objective that they would all leave the land and never come back. Jimmy Junior, the eldest, served his time as joiner, as did Albert, the youngest. Joe became the fastest plumber in the North-east. His father told me proudly about Joe putting in a bathroom in a new house and the main contractor arriving to find no sign of the plumber who had promised to come to his job next. 'That bugger o' a plumber nae here yet?' he fumed at his foreman. 'Oh aye, he's awa aaready – finished.' Dod did a variety of jobs, including driving one of Crighton's (the Methlick baker) vans, where his father was very proud of Dod's reputation for selling more than any of the other vanmen. Dod later sold insurance and drove buses but he never did, perhaps he never dared, try the land. And Jimmy was pleased that Belle, his only daughter, married a master-joiner and so was never 'cottared' (the somewhat pejorative term for living in a tied house on a farm). So it was no wonder that he thought little of my decision to give up a senior lecturer's job. 'Fut the hell are ye deain geain up a gweed job for fairmin? If ye wis gaun tae dae that you should have bidden at hame and learned tae fairm richt.'

And yet I got off to the best possible start. One of the things the little farm needed was more acres if it was to be a 'richt place' again, and in my search for a house I was able to kill two birds with one stone. Backhill of Ardo Croft, just over Ardo Hill to the north, included twenty-one acres of land which had been poorly farmed for many years. The Patersons, who had a long-established re-putation as millwrights, had no further need of the place. There were still several joiners in the family but they were getting old. The house, which was in very good heart, was totally out of date (it had no bathroom but did have an outside water closet with a fine view up Ythanvale to the Braes o' Gight) and the steading was falling down, though I did rather like the warm red of the rusty corrugated iron roofs. I was able to get an 'amalgamation grant' so I was able to buy Backhill for under £3,000. It was a small step but one in a direction in which the farm so badly needed to go. And because we had the old Smiddy Croft beyond, it consolidated the farm as well as adding to its extent. We had bought our large Victorian mansion in Glasgow for £4,000 and we now sold it for £20,000. By the time we had spent all that profit on doing up and extending Backhill we had a really good house that would do us very well until the farmhouse became available.

When Backhill was being done up preparatory to our move north, we got a warning of little tensions that could arise. Tawse Brothers of Methlick, who were the main contactors, had a cheeky chappy of a joiner and he tried some lip on James Low from the safety of the roof while he was putting on the sarking. 'Aye Low,' Norman Stott shouted, 'How are ye gaun tae like it fan Charlie comes hame and ye've twa bosses telling you fit tae dae?' As usual

James Low was quick enough. 'I really couldna say,' he replied, 'for I hinna had nane up till now.'

For the new farmer, it was a big change from the university. You met a different class of person on Ardo Hill, especially the heroes who clung to the small farms to the north of us on acreages that were too small and so necessitated great ingenuity on their part. Like the Beatons of Brownhill Farm, who fed calves, which required a small initial outlay, and could yield a good profit if sold at the right time and looking well in the store rings at Maud. They had free range eggs before anyone thought of calling them that, and half an acre of delicious Golden Wonder tatties which could be bought by the bag or by the stone.

John Beaton, the farmer of Brownhill's seventy-three acres, humbled me greatly when I went, the farmer of the grand place to the south, to buy a load of turnips for the cattle which were selling for as much each as John would have admitted to making in a year. The neeps were duly pulled by hand and loaded, shaws and all. Now how much would that be?

'Oh, one-fifty'll dae fine, Charlie, seein's it's you.'

The depth of the ignorance of the latest farmer of Little Ardo was such that I didn't know whether John meant £150, in which case I would be expected to argue the toss, not easy once the neeps are in the cart, or 150 pence. 'Seein's it's you' could have meant 'That's a bargain as you're a fine chap and neighbour with whom I used to play football', or it could have meant 'Seeing you're selling cattle for £5,000 apiece and I have to make a good job to get a hunder poun for mine.' The decent man, of course, meant £1.50, for perhaps two tonnes of beautiful Route Tofte turnips.

Then there was Bertie Taylor, who farmed next door to John Beaton at Cairnorrie Farm. Bertie was a good farmer who made most of his money from the short keep store cattle market. He would buy at Maud market when the cattle looked cheap in relation to the availability of keep, and sell them wherever he saw a profit. The profit didn't have to be great and the wait didn't have to be long, but there was usually a profit. Unusually in Aberdeenshire at that time, Bertie also liked a deal among horses though I was never sure whether that was just to please his children, who were the envy of their contemporaries because there was always a shelt about Cairnorrie Farm.

Bertie was also remarkable for his ability to make hay. Aberdeenshire farmers have always been bad at hay. The advent of silage as a method of fodder preservation, which only really got going after Hitler's war, was a great thing for the North-east. Our fathers grew the wrong grasses for hay, and they never had the patience to wait until the haar had retreated far enough back towards the North Sea before rushing to turn it, which was much worse than useless unless it was going to be ready that day for coling (or, in later years, baling).

But no matter what kind of a year it was, no matter how bad the weather, Bertie always made passable hay. And if he ever got good weather, he could make hay that was almost as good as that gorgeous stuff they produce every year in the Carse of Stirling – and a lot cheaper.

When the new farmer of Little Ardo went in 1973 to buy a few bales from Bertie he got some grand stuff for £1 per small square bale. He could have had little profit at that but Bertie was pleased enough with a pound. And when the oil and the difficult farming times in the late '80s brought the flood of White Settlers to the North-east, Bertie enjoyed good times with the haymaking. The horse population of Aberdeenshire was multiplied severalfold, as parents bribed their children into accepting the move to Scotland with the promise of their own pony. Bertie was soon up to £4 a bale . . . but that was only for the settlers. We old-established Buchan farmers still got our few bales at £1 a bale. In those highly inflationary years, when the price of everything went up every year (as much a 26 per cent in 1977, and always at least 3 per cent), Bertie kept his local price steady. We who are slow to give credit gave him a lot of credit for that.

The farmer of Cairnorrie Farm had yet more success with his organic oats. One of the things that the White Settlers reinforced in Aberdeenshire was the organic movement. It was thought by enthusiasts to be healthier to eat porridge that had been grown with inorganic weed killers, fungicides and manure, or none. And those who were converting their crofts to organic status needed the oats they fed to the inevitable ponies to be organic oats.

Bertie was one of the first to spy this as a window of opportunity. If he could get a premium for selling hay to the horsey folk he could make even more growing organic oats for the foodies. This led to a most interesting conversation between the farmers of Little Ardo and Cairnorrie Farm. In 1991 Bertie had grown a fine stand of oats in the ten-acre field beside the Methlick to New Deer road. Oats were not that common by this time in Aberdeenshire, so everyone had a good look, especially on Wednesdays when they were driving to or from the great markets which were still held at Maud. You couldn't help noticing that Bertie's organic oats had regular stripes about thirty feet wide, of luxuriant oats and slightly paler, thinner shorter-strawed plants, time about, right across the field. As spring turned to summer the stripes grew more and more pronounced.

Those stripes were a bad sign. Bertie was needing to sort his sprayer, which was spraying unevenly, or else he was spraying when it was too windy. Everyone took a pop at Bertie at the mart. 'Aye, Bert, I see you're usin' the manure oot o' the strippit bag' . . . a good farmer like Bertie doesn't like to have his mistakes so near the road, but Cairnorrie Farm was all like that, for it straddled the main road.

I too had a go at Bertie but I wanted to tease him, not about his machinery or his spraying technique, for I had little room to speak there, but about the organic status of his oats. It was quite clear that he had been putting on inorganic nitrogen, and plenty of it.

'Aye, Bert, that's nay stripes I see in yer organic oats, surely?'

'A wee bittie, maybe,' said Bertie with a slight shy smile, for he was not sure yet of the direction of the attack.

'But you're nae supposed to put on nitre on organic oats,' said I, just a wee bit self-righteously.

Bertie was most dismissive, 'Ye canna grow corn without nitre,' he exploded, with some reason.

It was only any of my business in so far as anybody's business is anybody else's in the country, and it always is, so I persevered. It was not through any regard whatsoever to what some White Settlers thought they should be getting or even that they should get what they were paying for, but I was interested.

'Bertie, if you pit on nitrogen on your crop, what's the difference between organic oats and ony ither oats?'

'Well, wi organic oats you've tae dae yer sprayin at nicht,' he said, as though nothing had ever been more obvious.

At Little Ardo we were surrounded by farms of about 100 acres, which were becoming far too small, but whose hardy owners hung on and even prospered by careful attention to detail in the care of livestock. Perhaps the champion was Billy Davidson, who reared hundreds of dairy calves on contract. He and his family made such a good job that on one year they got national publicity for rearing a thousand calves with only one death. That was exactly ninety-nine fewer than pessimists warned you to expect. Jimmy Bruce and his eponymous son, our nearest, in the sense of most in-our-faces, neighbour had a most un-Buchan-like flock of beautiful Border Leicester sheep. As typical Buchan farmers we had no idea what they were doing, but they were always among their sheep and you didn't need to be a shepherd to see that they were doing them well. Arthur Glennie was at Hillhead of Ardo, where he did better than any of us by fathering Evelyn Glennie, the virtuoso percussionist, but not far short was the thrifty job he made of his mixture of sheep and cattle. Arthur played the accordion, and I like to think that his wonderfully gifted daughter had heard a good deal of that, and that it was always there for her in her head after she went deaf. Then there was Jock Paterson, farmer of Gowkstone and my boyhood boozing pal, whose father had also had Newseat of Ardo and had survived with the help of speculation on the stock exchange. Jock, who had, like his father, seen the hopelessness of the smallholder, served his time as a mason and became the ace slater of the district. But he kept on his ten cows until Anno Domini closed him down in 2004.

Those are only a selection of our many neighbours, with the

same problem of size as Little Ardo had – only more so. They were tough guys, clinging on to their heritage. They embody the spirit that made Buchan, a meagre gift of nature, the productive land that it is. There is one image I would like you to hold of the peasant and his labour of duty and of love. As his nineties approached and the two replacement hips he had been given in his seventies started seriously to deteriorate, Gordon Thompson of Touxhill still went to the field to prepare neeps for the byre. With his tapner under his arm he struggled down the drill with his two sticks. Then when he got to where he wanted to start he threw the sticks, one at a time, down the drill. Slowly he bent himself double and lifted a neep, chopped off its roots, then its shaws and put it down. Without rising he then struggled a few inches to the next neep and did the same again. When he eventually reached his sticks, he retrieved them and used them to stand up straight and stretch. He then threw them farther down the park, slowly bent himself double and started again. Yes, my neighbours were tough guys, with a much better grasp of what they were doing than the lad who came home to Little Ardo in 1975.

I was in Glasgow when the great day came that we would send a couple of our foreign cows to the Highland Show. This meant 'You twa buggers', as James now came to call his new boss and his man Willie Adie, had a lot of new doors to go through. Willie and I had herded cattle but never led them on a halter. In fact I had never even put a halter on a beast of any kind at that stage. We had scraped the sharn off their tails and washed their udders at milking time but we had never given our cows a shampoo and blowdry. Still, we importers were all expected to do our bit to promote the new breed and so we had to learn showmanship. Sadly there was no class we could attend – the only way to learn was at the school of hard, expensive and embarrassing knocks.

Tulipe was certainly the showiest heifer we had. Her French owners had gone to a great deal of trouble to train her horns so that they curled round like a blackface ram's. So the first job was to get a halter on her and see if we could hold on. We tied on twenty feet of extra rope so we could both hang on, and to give us more chance of getting it round a post or a gate when, as we were sure she would, Tulipe would make a bolt for freedom. We got her into the crush, got the halter on and then let her out into the midden. We thought it was fortunate that there wasn't much muck in the midden at the time and, there being a high cement wall around it, Tulipe would not escape whatever sort of rodeo developed.

Well, we needn't have worried about the rodeo. Just as we hadn't a clue what we were trying to do, neither did Tulipe have any idea what these idiots who didn't even speak French wanted her to do. Me pulling and Willie behind prodding and then twisting her tail, we tried to get Tulipe to walk for the judges. She simply would not budge. This would never do for the Highland Show.

To our humiliation, although ultimately to our great benefit, there was just one judge watching this first attempt at showmanship. James Low was leaning over the midden dyke watching. His pipe seemed to be even sweeter that usual. I thought he looked a bit smug, but really it is a wonder he didn't burst into peals of laughter. It is one of the scenes that I wish somebody had been filming for posterity. At any rate, when he had seen enough of this pulling and pushing and twisting and prodding, James Low, a man who had had the Horseman's Word and knew all about halters and how to lead a beast, came down into the midden.

'See hudd o' that tow,' which would translate to 'Would you mind awfully if I had a go?' The first thing he did was to remove the extra twenty feet of rope with much effort, for the rope was by this time very tight indeed. He then took a very short hold of the halter, scratched her ears for no more than three seconds, made that kissing sound with his lips that used to tell a good horse to move forward, and off Tulipe set at Jimmy's side. When he came to the other side of the midden he just turned her round, with no tugging or tearing at all, and walked her back. When he reached, 'You twa buggers' he led her in a tight little figure of eight for a good minute and neither of them was even sweating. Then he left us to it.

We took two heifers to the Highland Show in 1972. After

Jimmy's lesson, the leading of them wasn't a problem. In fact, by the time Willie led Tulipe at the Edinburgh he was instructing her by word of mouth rather than hauling on the halter. 'Walk on,' 'Stop,' 'Stand', he would say and, good girl that she was, Tulipe did just as she was told. Both the heifers had been led at home, one in France and the other in Germany, so that was no problem after the cattlemen had got a clue. But we didn't feed them properly for a show. Little Ardo had always been a commercial farm where cattle were kept fit rather than fat. Fat cows don't calve easily and they tend to put their food onto their backs rather than into their udders. Both our heifers were too thin. There were ten in their class and they stood ninth and tenth. No one said a good word about our cattle. And that shows just what a lot of rubbish the show ring really is. It tells us nothing about the quality of the animals. What it shows is the quality of the stockman, and the effort that has been put into preparing them for the show. I can say that with some confidence because, with the right feeding, I was able to bring Tulipe back to Edinburgh some four years later, when she won the Championship for me. Gertrude's first calf won the junior bull class at the Royal Highland Show two years later and George Anderson of Keir, who bought Gertrude's third calf, won the Breed Championship at the Highland and a string of other honours with her in 1988. The judge who placed her Supreme Champion in 1980 said of Tulipe, 'Where have you been hiding that one, Charlie?' and yet he had seen her at her first show. It was certainly the cattlemen who were last in that class of ten in 1972.

Winning things at the shows, with all the work involved in feeding the competitors, halter-training them, shampooing them, clipping them and giving them endless blowdries has very little to do with farming. It has nothing to do with the production of good meat, good milk or good leather, plenty of it or economically. There are prizes to be won, but the expense of putting a team to the Highland Show with the cattleman's maintenance to pay, could not, in those days, be justified by the prize money – even if you won everything. Showing is really most like sport. It is about competition. It is about winning or even about losing with a good grace. One of the finest pedigree cattlemen in Scotland is Jim Jeffries of Kerseknowe at Kelso in the Borders. He sold the great Charolais bull Kersknowe Festival, described by Archie Whyte – proprietor at the time of the Spott herd of Aberdeen Angus, one of the founding herds of that opposing breed – as 'the finest bull of any breed I have ever seen'. Jim judged often, but he didn't show at the Highland or even at his local shows.

It is true, for the top herds, that there is some commercial advantage to successful showing. A win at the Highland will be remembered when your bulls come into the sale ring at Perth. And if you have distinguished yourself in the show ring people who are looking for replacement heifers will bear your herd in mind and you may get an extra couple of hundred for bulls sold from the farm, without the trauchle of bringing them out for a public sale.

But it is the sales that count. The great Bull Sales, which were held at Macdonald Fraser's mart in Perth every October and February were, until their move in 2009 to Stirling, where a pedigree breeder of bulls looked for a large part of his living. The new farmer of Little Ardo knew that, while he hoped the Americans would soon be back to buy his calves, in the long run he was going to have to take bulls to Perth and sell them well if the cattle trade was going to keep him in the style to which he aspired and restore Little Ardo to its former glory.

Our first foray to Perth was with Charolais. We had a decent bull. We had given him plenty to eat but we had so much still to learn. His hair wasn't long enough, thick enough or strong enough, which was down to the fact that our bull shed had been built to keep pigs warm, whereas a bull should be as cold as possible. If cattle are cosy in their stall they will moult. And why is hair so important? We don't eat it and the bull's hide will make just as good leather with a thin coating of hair as with a thick one. Well, hair is important in the business of making poor bulls look not too bad and good bulls look like great ones. We were taught that lesson at Perth when the cattleman from Asloun in Aberdeenshire gave us a hand to put out a Charolais heifer. Willie and I had done all we could and the heifer was looking fair, at best. Kenny Stewart couldn't bear it. He came over with his comb and in fewer than thirty seconds he had the beast looking at least a hundredweight heavier, with her hair all standing up on end. From last she was up to average by the time she entered the ring. She made 2,200 guineas and I'm sure Kenny's thirty seconds was worth 500 of them. And the lesson was worth more. The curry comb we had used for getting hard sharn off dairy cows' tails was not what was required. A broad comb with a few long and thin teeth had to be drawn in big sweeps over the animal's back. The little dabs that Willie and I had been employing were quite useless. We were like crows with a new road kill when they don't know quite where to start.

So thanks to Kenny we had done all right with the heifer, but what we were really there about was selling our first Charolais bull. He was fat enough, but he had shed all his hair, so nothing could be done to make him look like a pedigree bull rather than an enormous pig.

The average price for bulls on that day was £2,100 and all we got for Ardo Hadrian was £600. If it wasn't the lowest price of the day it was pretty close. And the humiliation wasn't over − not nearly. The buyer, a very tough character from the north of England, came and demanded a 'BEET of LOOK' which I eventually twigged was a 'bit of luck'. He wanted a few quid back in cash to help him cheat the taxman. I knew that paying something was a good idea if you wanted your buyer to buy from you in future, so despite the poor price I gave him a fiver. I should have told him to enjoy the bargain he had got and to bugger off, but I didn't. And he wasn't satisfied. He could see I hadn't a clue what I was doing, so he now asked a fiver for his son and then one for his cattleman. Oh dear!

On the way home I cheered Willie and myself up with a lot of wrong-headed rubbish about how £600 was the price of three fat stots and they always said a bull was worth five fat stots, so we were only two short . . . and all those guineas for the heifer would come in very handy. But it was daft to try to sneak that one past Jimmy Low. When I told him we had only got £600 for our bull at Perth he snorted in perfect disgust. Pluckily I offered 'Well, it's a start.'

'Aye,' said the grieve with an extra helping of his usual charm, 'a bad een.' The worst of it was – as usual, he was right.

So Jimmy Low wasn't impressed, but there was a lot of interest in the countryside, in this lad giving up a good job at the university to come and start a menagerie of foreign cattle at Little Ardo, where nothing had ever happened since Maitland Mackie had had it. Young farmers came to hold stock judging competitions to see if they could grade these fancy cattle. It seemed like every farmers' discussion group in Aberdeenshire, and some from as far south as Forfar, either got me to speak to them at one of their evening meetings or wanted to come for a farm walk. Those were good fun if a bit embarrassing, especially for our two sons, who were now at the very rural Turriff Academy. When they heard that the Turriff Discussion Group were coming the boys pleaded with me not to let them see the tractors. Their pals had been quite impressed that we had two Fords at Little Ardo but the thought of these lads finding out that Little Ardo only had two Fordson Majors caused panic. The boys weren't altogether wrong, for the farm walk is not just a chance to admire. One of its most appealing features is the chance it offers, after the admiring noises have been made and the vote of thanks given, to criticise what they have seen on their way home and with everyone they meet.

We had many such educational occasions at Little Ardo in the early days of the exotic cattle, and in particular I must tell you about a wonderful night with the Auchnagatt Discussion Group. I had had the usual problems with the subterfuge of the farm walk. One of the points of interest was my barley beef shed. It was to double as the venue for the discussion if it wasn't a fine night, so we had to hide the eighteen bulls who had pneumonia in the Dutch barn. That was a pity, as it meant I couldn't show it off to sufficient advantage and it was the tallest structure there had ever been on the farm of my ancestors. We had even got a skip and loaded all the old metal that over the years Mr McConnachie had rejected as valueless even as scrap on his annual recycling tours. The old tyres we used to put on the silage pits to keep the covers from blowing away and a year's supply of black plastic also went on the skip, so improvements had been made. It was a fine night and we settled down on a ring of bales in front of the old farmhouse to discuss the ferlies they had seen, lubricated by the beer that Peter Crighton the baker somehow managed to get for us at one shilling a bottle. For many of my guests it was the first time they had seen Gelbvieh, Chianina, Romagnola and Blonde D'Aquitaine cattle, so they were interested, though they were mostly interested in what the crosses would be like. No one knew, but that didn't stop a good argument getting up. Soon the sandwiches and the beer were gone and everybody wanted the whisky of which I had laid in two bottles. I had brought a few glasses but not nearly enough. My parents were still in the farmhouse but were away on holiday so I couldn't get any from there. Then Gordon Paterson, my resourceful grieve, had an idea. In a minute he was back with the calfies' pails quite well swilled out. Two of my guests were soon away to the village for more bottles of whisky. As I wrote in my diary, 'By nine o'clock we had exhausted our interest in cattle. By ten o'clock we had got fed up of agriculture and sorted out the national and international political scene. By midnight the good people in the village below were being entertained to the most enthusiastic renderings of the old ballads carried down to them with the evening dew. They did all get home safely, but Albert Howie (the very jolly farmer who later became a crackshot of the Limousin and Blue de Maine sheep trade), went home wearing his natty Italian shoe on one foot and a potato crisp packet on the other. We found the other shoe about ten feet up on the big beech tree my father and I had planted in 1946.'

CHAPTER FIFTEEN

'Wembolie, Wembolie, Wembolie'

1978

One of the many social highlights of our life during my first attempt to be a farmer was the trip to Wembley stadium to see Scotland play England at football. In the normal run of things the last thing I would have chosen to do with a spring weekend was to join the great unwashed on the train to London to watch a football match. Besides, my parents, who had spent some time in London and my aunts, who lived there for many years, had told me often of their shame at their fellow countrymen causing mayhem in the capital every second year, fighting in Soho and spewing in the tube trains. But in 1978 it was a matter of principle that I should go. Ted Croker, the secretary of the English Football Association, forbade me from going. No tickets were to be sold through the usual network of Scottish clubs. All the tickets were to be allocated to English clubs. This was not pure English chauvinism, though there is no doubt there was a bit of that. The main reason, and certainly the official reason for banning the Scots, was that two years previously Scotland had not only beaten their hosts but they had demolished much of the stadium. They had dismantled the goalposts and taken them home to Scotland. They had dug up great divots of turf and taken them home to Glasgow and Inverness. And it wasn't just the neds who took part in the dismantling of the headquarters of English football. My friend, former colleague from Strathclyde University and the future Lord Provost of Glasgow, Sir Michael Kelly, Dr Mike Kelly as he still was then, was very proud of the patch of Wembley turf that graced his lawn in Pollokshields. And it wasn't on the drying green round the back. Mike planted his little piece of football history in his front lawn.

Ted Croker was not for any repeat of that performance, so he would ban the Scots. It was an ignorant move. If he had wanted to be sure of selling all the tickets for the match twice over he could not have devised a better marketing strategy. In no time Scotland was awash with tickets. Everybody wanted them. I got two from Al Dalton, an oil man who lived in the village . . . two £3 tickets for £64, a price I deemed a bargain. No Croker was going to stop me going to a football match in my own free country – even if I didn't want to go. We had taken our eldest two children, Sarah and John, to London to see the sights, so I would take the next one, Jay, and show him the capital. The trip would be educational as well as sporting, and that would show Mr Croker.

We took a sleeper from Aberdeen, resisted the temptation to join in the singing and boozing in the buffet car, though there was no getting away from the heavy traffic back and fore from the compartments to the bar. All the same we slept reasonably well and arrived quite fresh at King's Cross. It was an ungodly hour, which was fine, for we had the sights to see.

What was also ungodly, or at least a bit unchristian, was the fact that Mr Croker had somehow managed to get all public transport taken off for the day, to make sure that no Scots who did make it to London could get to the stadium. There were no buses. The tubes weren't running, so we set out on foot to see Buckingham Palace, the Houses of Parliament, to get our photographs taken at the door to Number 10 Downing Street, to which there was open access in those days. There was nothing very memorable about that part of our trip except the taxi driver who stopped and in a strange reversal of roles, hailed us. ''Ere,' he said, 'I just want you to know that I think it's a bleedin shame what they done to you people. You come down 'ere for the match and there ain't no transport. It's a disgrace.' So it was. We couldn't agree more, but then he just drove off and left us on the embankment beside Cleopatra's needle.

He wasn't allowed to give us a lift so he didn't, however big a disgrace to the capital he might think it.

We made it to Trafalgar Square just before nine o'clock. The cultural bit of our trip was over. Now for the fun. And it really was fun. There were thousands of young Scots in the square. Most were more or less sober, though they had been down all night and many had only a tartan tam-o'-shanter, tee shirt and a saltire or a lion rampant on. You could hardly see the fountain for all the boys swimming, splashing and climbing as high as they could get. It was very good-humoured. There were very few policemen there so they didn't try to keep order. I managed to keep Jay out of the fountain, but only just.

About half past ten a consensus developed that, good fun as it was in Trafalgar Square, with eleven miles to walk, and only four and a half hours to go till kick-off, it was time to move off.

So off we moved, a seething, jolly, noisy mass. A human tide flowing towards Wembley singing 'Here we go, here we go, here we go', 'Que sera sera, we're going to Wembolie' and, 'We're on our way to Wembolie, we shall not be moved.' It was very odd really, for whether they were supporting Mr Croker or whether they were scared of the tartan hordes, I don't know, but there were few shops and no pubs open. No traffic, and certainly no buses and no taxis. Having abandoned Trafalgar Square to the pigeons we briefly took over Piccadilly Circus – mobbed it. Then one of the pleasantest moments in a very odd occasion. As we made our way down Regent Street, in the magnificent terrace that curves out from the Circus, there was one shop that was open. It was a travel agent's, and sitting at her desk trying to be busy was a very pretty girl with nice legs. And she was being distracted by a male voice choir of maybe twenty lads singing at the tops of their voices, the British Airways advertising jingle of the time, 'We'll take more care of you. Fly the flag. Fly the flag'. She enjoyed that, so the lads then sang 'Isn't she lovely? Isn't she bee-u-ti-fu-ul? Isn't she lovely, fit for love?' She certainly loved that and I am sure there is at least one middle-aged lady who still looks back with affection on that tartan army.

If only there had been a pipe band it would have been perfect, for we were swept along on a tide of laughter and song and jokes about Mr Croker and what we would do to him if we happened to see him at the game. But it was a long way. There was no food to be bought and the cairry-oots soon went done. The flimsy clothing, and feet soaking from the fountains of Trafalgar Square meant many of the troops were cold and uncomfortable even before the drizzle got going and the hangovers set in. The mood came down off its high and stoicism was the order of things, but the essential spirit remained. I will never forget this little keelie walking along, huddling his saltire round his shilpit shoulders, his soaking See-You-Jimmy bonnet with the ginger hair, and he was muttering to himself, 'I'm gaun tae fuckin Wembolie. Ted Croker'll no stoap me gaun tae fuckin Wembolie but.' Just as Jay and I were overtaking him, we were all overtaken by a mounted policeman. He was magnificent on his great chestnut gelding, and comfortable in his police cape. When the wee keelie saw him he stopped, looked straight up at him and, pointing his finger at the policeman, sang out 'Robin Hood, Robin Hood, riding through the glen, Robin Hood, Robin Hood whaur's yer band o' men?'

The policeman was not amused. He was too important for that. 'Oi,' he snarled in a really unpleasant copper's voice, 'you wouldn't come over 'ere and say that.' He could hardly have been more wrong. The words had not died in his throat before he was surrounded by a dozen or fifteen lads all excitedly pointing in time to a ringing chorus of 'Robin Hood, Robin Hood, riding through the glen'. The bobby was wrong, but he was also an idiot. He could have given the lads a smile and wave or even a clap, but no: he just threw in his hand, dug his heels into the gelding's flank and galloped off, utterly vanquished by good humour and followed by a derisive cheer.

We trudged along and along. My feet were sore but at least I wasn't wearing a tee shirt, a saltire and a hangover. Jay seemed quite buoyed up the whole way, though he had stopped talking by the time we had been walking for three hours. As we got nearer and nearer to the stadium the crowds, which were descending from all directions, got thicker and thicker. I remembered the lesson the big boys had taught me when we used to go to Pittodrie; the way to progress in a crowded street is to look for gaps and jump into them. There you could walk without tripping up or kicking others' heels. Whenever we saw where the crowd was a little thinner we were in there. But it was a long walk and there was no pipe band.

The eleventh mile to the stadium was really slow but the Glasgow humour never failed us. In my memory he is indistinguishable from the keelie in the story about Robin Hood. He was small. He had a tee shirt on and wrapped around him was a saltire under his See-You-Jimmy Kilmarnock bonnet with the straggly red hair dripping down his neck, and he was mumbling about his determination to get to the match. 'They'll no stoap me!' Then nosing through the crowd at maybe four miles an hour glided a huge white limousine with darkened windows. The crowds parted. To my astonishment no one did anything to even the situation up – they just moved to the side to let him through. But our hero didn't even see the car coming up behind him. He just trudged on oblivious and the driver had more sense than to blast his horn – or even give him a gentle toot, for football crowds can be excitable. So the wee man from Glasgow seemed to be leading the car. It was like an old-fashioned steam engine with the obligatory flag-carrier in front. Eventually Wee Jimmy became aware that the crowds had parted and he had a big share of the road to himself. He turned round and saw the beautiful white car. His little face, which had been screwed up against the cold and the pain and set in determination to get to Wembolie, gradually opened up. He was smiling. He opened his arms ready for a huge embrace. 'Daaady,' he roared, 'you've came for me. I kent you wouldna went aff and left me here.'

There was more good humour when we arrived at Wembley at about half past one – an hour and a half to go to the great kick-off. In the huge and still-empty car park there was a 150-a-side football match between Scotland and England. That was a vigorous affair but there was no fighting. In fact it was the first time I had seen a noticeable number of English fans. We watched for a while and I

181

managed to keep hold of Jay to stop him joining in. It would not have been as much fun as it sounds because in 150-a-side it is not easy to get into the match. I watched one tubby Scot who was eager to be a part of the first Scottish victory of the day, but in the ten minutes he just trotted towards the ball, like a piece of flotsam in the shallows, as the waves surge back and fore, and he never once got a kick, not even at an Englishman.

Now, Mr Croker might not have wanted the Scots to get tickets for the great game, but there was one group of Englishmen who were very keen indeed that we should have tickets. We had several times been offered tickets in the street. In fact we could have had them for £25 and saved me £14, for the touts were everywhere. There was a good deal of banter on the long march. 'No no. There's plen'y tickets. We'll ge' them for a tenner a' the grund.' They are some boys! Imagine going all the way to London for a football match and trusting that you'll be able to get a ticket at the match when selling at the ground is illegal!

At any rate, the time came when business had to be done. There was a line-up of touts against a wall at the ground. They had handfuls of tickets but they were not finding them easy to cash. There would have been twenty touts there and perhaps two hundred young lads who needed tickets. But young though they were, these Scots were street-wise. They had seen the asking price of these tickets come down from £50 to £15. They had been to Wembley on demolition duty two and four years ago and knew that for Wembley this was not a big crowd. Nobody was buying. 'Hey Jimmy, I'll gie ye a tenner for yer tickets and let you away hame, aye, a tenner for the lot.' That was maybe the best offer he got. We don't know, for Jay and I felt it was time to take our expensive seats. When we left the excited marketplace 200 lads were giving the touts a rousing chorus of 'Your tickets are fuck all, Your tickets are fuck all, your tickets are, they're really worth fuck all', to the tune of 'Those were the days my friend'.

Mr Croker must have been disappointed when he got to his seat. But when we got to our £32 seats we were delighted to see that the stadium was only about four-fifths full and that at least two-thirds of those in attendance were supporting the visitors. Far from the Scots being kept out, they were there in record numbers. It was the English who didn't turn up. Many had sold their tickets to idiots like me who were willing to pay anything to show Mr Croker where to get off. Some must have been put off by the lack of transport to the ground. But I like to think that a few English fans didn't show up because they found the whole business of banning the Scots distasteful.

There was also at least one distasteful joke. It was the time that poor Bobby Sands was becoming a nationalist martyr in Northern Ireland by starving himself to death in the Maze prison and one lot had a banner that read 'We've even got a ticket for you Bobby Sands.' That was a bit too much for me.

Soon after we got to our seats we found out how the confrontation between the lads and the touts ended up. As the match drew ever closer the sellers had grown more and more anxious and the buyers had grown more and more cocky. Then, with ten minutes to go, the touts threw in their hands. They could see their tickets were not going to sell. They could see dangers in trying to take the tickets away with them, and by half past four they would be worthless anyway, so they threw the briefs at the crowd and slunk off. Our informant was one of the lucky ones. He was able to pick up his £3 ticket for nothing and he landed next to Jay and me in our £32 seats.

And what about the big game itself? Well, it was not a vintage match like Baxter's game, where Scotland thrashed the World Cup winners 3–2 and Slim Jim was so laid back he actually played keepie-uppie and even had time to sit on the ball. This was just a dour struggle in which every English attack seemed to end with Willie Miller emerging with the ball. Near the end Stevie Archibald, Scotland's wispy red-haired centre-forward, burst through the middle. He was clean through when he was brought down. It was a professional foul and John Robertson tucked the ball low into the left-hand corner from the penalty. With Willie Miller completely dominant at the back there was no danger now of Scotland losing. Of course, as my father used to say, 'The bite's nae yours 'til it's in your mou,' But soon it was and we chewed away in triumph. There was only the one goal and we were left, still with no public transport, to hoof it back into London, singing 'there's only one Willie Miller'. No one seemed to have taken the goalposts with them this time and we never saw any act of vandalism or violence. I think for the football-supporting Scots that was an unspoken undertaking. Croker was wrong to try to stop us coming, but the Scots' conduct in the past had left room for improvement. That unspoken covenant of good behaviour was illustrated by a wee nyaff striding along pulling over the traffic cones which were there for his safety. Nothing was said, but a mannie with a flat cap just came along ten yards behind him setting the cones up again. There might have been more violence if the opposition had turned up . . . but they hadn't.

CHAPTER SIXTEEN

A Bonnet Laird

1977

My ambitions to redd up the old place and heave it up to date were jeopardised by the collapse of the exotic boom before I even got to the farm. But still, I think it might have been possible had it not been for two things. The first was the explosion of interest rates in the mid 1970s. As I write this in 2010 my farming friends are paying 4 per cent or less for their overdrafts. But I never got money at below 10 per cent and within a very short time bank rate had risen to 15 per cent. It peaked at 18 per cent and I was paying 3 per cent above base rate. When you add the fact that my interest went on monthly and compounded, the annual percentage rate on my debts peaked at as near as damn it 23 per cent. Despite that, I built three new sheds, put up college-designed (that means designed by someone who had never done anything) handling pens, put down almost an acre of concrete on closes and roads, spent all the profits we had on our various housing moves on modernising the farmhouse, built up to 700 head of cattle and upgraded the machinery from 'scrap' to 'outdated' – an expensive advance.

But big as the problems of the cost of borrowing were, the second reason that farming became so difficult for me was that I had hardly started the job when my parents decided they could stand it no longer and they wanted their money out. With far too many commitments and far too little cash I was suddenly asked to find another £100,000 so that the old folks could enjoy the retirement they deserved – in peace, away from the infernal bustle their son was causing at the old place.

Indeed, it was soon all change at Little Ardo. James Low, after forty-six years in the employ of my father and grandfather, and thirty years sharing the farmhouse with John Allan, could see that it was time to move on. He got the chance of a council house in Ellon and suddenly he was gone. My father gave him the gold-handled cane my grandfather, whom Low had admired beyond anyone, had got when he retired as a director of Lawsons of Dyce. The village was never the same without him taking everyone and everything down to size with the style which was all his own. But it wasn't the end of his character. In no time he was grieve of a new gang of old men in Ellon, where he continued taking the country through hand. He always said that the Knockothie milk, which he was reduced to at Ellon, was thin stuff, and he endeared himself to many by shouting in the street at the proprietor of that dairy, 'Aye Davidson, if ye haud a suppy cake at that coos, yer milk winna taste sae like water.' It didn't endear him to Sandy Davidson.

Putting the old house together again had been done by my parents, who had a couple of years enjoying the luxury of occupying the whole house. But that only involved opening the door which had separated the grieve's and the farmer's dwellings for thirty years. Before we moved in we spent the money we had made off our Glasgow mansion and reinvested in doing up Backhill, on modernising the farmhouse. Once again Peter Murchie, our long-time achitect friend from Dundee, helped immeasurably, especially when he forgot to put in his bill. He drew us a third new kitchen, an extra bathroom, a secret passageway from the farm office so that the farmer could escape creditors, and brought two little rooms and a cupboard into what had been the old farm kitchen to make a really spacious lounge suitable for a big farmer with pretensions.

In 1977 Jean and John left the home to which the Hero had returned after Hitler's war.

They bought a mews cottage in Lincoln where my father said they had all they needed. Crucially, they were within easy reach of my mother's two sisters who were their favourite people. That was what John Allan said, but I think he was lying. He had written that while people might rave about the deserted beauties of the Highlands, or the architectural wonders of the Backs at Cambridge, he must have fields with crops and beasts in them if he was to be at peace. He needed to be in an active Lowland farming area, which Cathedral Street in Lincoln was not.

As well as wanting to get peace from my attempts to modernise the place, he left because he knew that one of the biggest drags on Scottish farming has always been the young men waiting for dead men's shoes. He believed it was part of the peasant's duty to his land to get to hell out of the way when he had done all the good he intended.

CHAPTER SEVENTEEN

Forced to Try Real Farming
1977–81

With the dreams of easy money flowing in from an endless line of Americans with more money than sense fast disappearing, the new farmer's main priority was to update the farming systems at Little Ardo.

The first bit of modernisation the new farmer had had to tackle when he arrived in the close in 1974 was the way the cattle were fed silage. The system that J. Low had employed for many years really gave the lie to my father's claim that by 1958, with the arrival at Little Ardo of the combine harvester, stupid labour had gone. Certainly the silage pits were now very handy at the neep shed door on the north side of the steading. That was grand if the silage was required in the byre. But what if it were needed to feed the beasts that were bedded in the Old Barn and went out to the midden to be fed? In that case silage was torn out of the pits by graip and thrown onto a cart. This was then driven round to a ridiculous construction called 'the New Neep Shed' which was tacked onto the corner created by the two sheds at the north-west corner of the midden. You never saw such a half-hung-tae thing in your life. Fortunately, in 1976 a gale from the south took pity on the old steading, got in at the open door and blew it away. But that was not before the New Neep Shed had been responsible for hours of stupid darg.

Once the silage had been driven round and couped in the New Neep Shed, it had to be graiped again into a barrow and wheeled along a cement trough and couped. The trough had no feed rail so a great deal of the silage landed on the ground and the cattleman was at the mercy of eager beasts that could easily cause him to lose his balance. I do think that the Health and Safety, mainly a pest, would have had a point if they had insisted that the poor cattleman got some support. A simple rail would also have stopped the cattle jumping up into the trough and making a mess.

John Allan had made improvements. The silage cutter, a thing like a medieval beheading axe which had to be swung high above the head to make a cut right across the face of the silage and so make a job which would have been almost impossible, almost possible, was replaced by a gango hammer attachment. It was like a pneumatic drill, only with a sharp blade on it rather than a chisel. With this you could cut across the packed silage with relatively little sweat.

The next improvement became possible when the back-acter was fixed to the power major. With this a determined tractor driver could charge the old Fordson in reverse into the silage and then, with much rearing of the tractor up on its back wheels, a decent forkful of silage could be ripped from the pit and the cart filled without manual labour.

This was still the system when I became the farmer. I 'improved' it still further. Instead of loading the stuff onto a cart and having to reverse it up the long narrow pass to the New Neep Shed, I had a brainwave. Jimmy Low called it 'the lazy man's way' which, with John Allan's training to despise useless labour, I took to be a compliment. Thus I rammed the back-acter into the silage pit, took out as big a forkful as I could and then took it straight round to the New Neep Shed without decanting it first into the cart. Our wee two-tonne carts would have taken perhaps six forksful at a time, but at least I didn't have to risk couping the cart into the midden while trying to back it into the dreadful neep shed's door.

It doesn't, I hope, take a time and motion expert to see that my latest improvement was no improvement at all. But the time and motion expert would have no way of knowing that my system had the added advantage of spilling a generous covering of silage all the way round from the pits to the New Neep Shed. It was not the obvious inefficiency of the system that led me to seek even better ways. The film of silage that covered the 200 yards from the pit was a source of annoyance to the three women who were the very foundation of the new farmer's slim chances of happiness. The trail passed Mrs Low's back door, which quickly became my mother's back door and soon became Fiona's back door. Ladies don't like mess at their back doors, especially if it serves no sensible purpose. At least the hens, those traditional polluters of farm doorsteps, did lay eggs.

I did have one success. I got rid of the twice-daily darg of filling and wheeling barrowfuls full of silage at hazard to limb if not to life. Instead of carting the silage round to the New Neep Shed, I made a feeding area in the south-east corner of the midden and dumped the silage there. It was fifty yards nearer and did away with the wheelbarrow, though not the graip, for the silage did have to be graiped in about to the feed barriers.

This was a great advance. But my new improvements weren't finished. Why not then move the feed barrier round to the silage pit and let the beasts go round and get it themselves? All I would have to do was move the feed barrier a bit farther into the pit every few days, or so I thought.

What was needed was an opening in the north wall of the only bit remaining of the steading which William Yull found when he came to Little Ardo in 1837, the Auld Byre. Now, there had been no door in that wall for the very good reason that several times a year the wind would come roaring down from the north and flatten every thing of which it could get hold. James Low was very unimpressed by his young boss's ideas of saving labour by making holes in the northern defences of the steading.

I knew he would not be pleased and took the coward's way of not telling Jimmy of my plans. Willie Adie and I just started to bash a big hole in the old stone and mortar wall. The grieve's reaction was a wonderful example of dour North-east scorn. When I met him in the close at lowsin time he said, 'I see you twa buggers have got a start to tak the reef aff the Auld Byre.' That was how the new farmer and his cattleman came to be known as 'You Twa Buggers'.

It wasn't a great plan. Scared by Jimmy's prediction that the roof would blow away, I had to rush round to shut the door and so block the hole every time the wind rose. That was always on a filthy night and the door was always blocked by at least a foot of dung and straw which was trailed out as the cattle went in and out. There wasn't enough drainage at the silage face to avoid a strang lochan gathering there. I gave it up when a valuable heifer somehow got herself caught in the feed barriers and, in trying to reverse out of this tricky situation pulled the feed barrier down on top of her. She nearly drowned.

Professional help was clearly needed to advise on how to make the best of the old place. And wasn't that what the College were for? When I told my cousin Maitland that Ronald Harrison, the College's buildings adviser, was coming he said 'Oh well. He's a bit pernickety but you'll end up with a good steading.' Mr Harrison fairly put Dad's ambition in context. He told me that Little Ardo was the only farm in Aberdeenshire that in his thirty years he had never been asked to advise.

To Harrison's plans we put in a braw new cattle-handling system with six covered pens. It was just like having our own mart. And this became the home for Scotland's first on-farm transplant service. To speed up the Klondyke of the exotic trade you could consign your cow to Little Ardo. We would prepare her for super-ovulation and cross-bred heifers into which to transplant the eggs. Then on the appointed day a firm of vets from the north of England came with a huge mobile operating theatre, retrieved the eggs from the cow and inserted them in the heifers. That was very profitable. We made money off hosting the deal, we made money off selling recipient heifers. Or at least we would have if the firm of vets hadn't gone broke. In the end we had to take our profit in successful transplants. Our most successful, an Austrian Simmental who had been Champion at Turriff Show, gave us eighteen fertile eggs and eventually seventeen pure-bred Simmental calves from the one operation. But by that time the bottom had fallen out of the exotic trade.

191

It had become clear that the exotic cattle weren't going to fill the gaps left by years of underfunding of the little farm on the hill. The trade never really recovered from the shock of the oil crisis of 1973. The Americans still wanted a few of the very best of the exotics, but only those who were very good at it would make money out of them in future. In 1973 we sold a Simmental heifer for 10,000 guineas and a very average little Gelbvieh bull for £6,000 to a Canadian. But having got £12,500 for the orra Chianina in 1973, four years later I sold the better heifer and her three daughters and two granddaughters by embryo transplant for £2,750 for the six. A few hardy men like Bob Adams of Newhouse of Glamis, Jim Goldie of Uplawmoor and Neil Massie of Blelak would always find a way by being at the top, but for the vast majority it was just an agreeable, sociable and even exciting diversion from the serious business of making a living on the land. Pedigree cattle were indeed, as many people had warned me, 'a rich man's hobby'. I had to get into real farming.

Stephen Mackie suggested I try Rodenight pigs. That's the system, called after the English farmer who invented the strategy of keeping your sows outside in little corrugated iron arcs. 'Delightfully reproductive' as Winston Churchill called them, the pigs would soon be producing as many as ten piglets each in an arc apiece. To keep the enterprise simple I would sell the piglets on to a fattener at sixty pounds' weight and start the cycle again. Each sow could be expected to produce two litters a year, no new buildings would be needed, I would soon be rich. By the greatest good luck Stephen, who had been rearing pigs in this idyllic system for many years, now had forty arcs and forty sows he could let me have at a very reasonable price while he moved back into more conventional factory farming methods with expensive farrowing sheds and specialist fattening houses. It never occurred to the novice farmer to wonder if what was right for the old goose might not also have been better for the young gosling.

In a couple of months the sows all pigged more or less at once with huge litters. There hadn't been a pig on Little Ardo since the war, so there were no porcine diseases about and deaths were

192

minimal. The mammas, having been over the course several times before, hardly lay on any of the tiny shivering morsels who all seemed to know how to tremble their way round to their mother's belly to fix onto a teat. Soon the Hill Park was carpeted by earnest little grunters who tore about in delightful waves of pink. This was surely the way to do it – and so wholesome. No do-gooder could object to such a wonderful system. It went so well that I was able to wean 10.3 piglets per sow, which I was led to believe was a record for the system, and that left enough money to pay for the original sows, all the arcs and pig feeders, and even to pay for another seventy gilts to add to the next batch, which would be all profit. Clearly this was the way forward.

Well, it wasn't. In another six months all 110 farrowed at once, and in a snowstorm. The seventy gilts hadn't a clue what to do. They pigged in twos and threes in the same arc which led to cannibalism and in the confusion there were many piglets overlain. It was now mid-January and the outside system which had seemed so good in July was a nightmare. There may not have been any disease the first time around but this lot were soon sneezing and scouring and dying in droves. They say that pigs only smell when they are losing money, and you know, I almost convinced myself that the first batch smelt quite agreeable. But the second batch soon stank horribly. There were no waves of eager little porkers haring about in an explosion of *joie de vivre*. Occasionally one would appear into the icy draught looking so lost you wondered if his mother had put him out. Soon I was taking a barrow when I fed the sows in the morning. I emptied the feed bags and then filled my barrow with the dead. The hopelessness of the situation was brought home to me when I got the vet. He cut the most pathetic figure I can remember seeing in my years in farming. It was a relatively fine day and when the sows came out to see if the commotion meant there was to be more food quite a lot of the walking wounded followed. The vet stood among perhaps 500 piglets looking miserable. Every now and then he caught one and sprayed antibiotic into its mouth, but he had no record of which had been dosed and could not tell me what was wrong or how to fix it. And neither could anyone else among my pig-breeding friends. Perhaps it was just too obvious to be worth mentioning, but I eventually worked it out for myself and I got antibiotic powder and mixed it with their feed. That quickly stopped the dysentery and the pneumonia but the damage had been done and the profit was gone. Four of the nine live births per litter were now dead.

And there's more. It turned out that those that had died lost less money than those that lived. The survivors' lungs and digestive systems were damaged. Their appetites were undiminished but their ability to grow had all but disappeared. They were eating £1,000 a week but they weren't getting any bigger. The stink of pigs became unbearable. The money ran out. The North-eastern's rep, Bill Allan (not a relation), offered to give me enough feed to see me through to another batch if I pledged my crop to him at harvest time. I was not tempted. Instead I weaned the lot early, sold the sows and the arcs and so raised enough money to feed out the rest. Well, not quite the rest. The last dozen I killed and buried. They would be with me yet if I had persisted in trying to finish them.

There was a farcical footnote to my first venture into pig-keeping. It happened when I sold off my sows at the fat sow ring at Kittybrewster mart in Aberdeen. As luck would have it, and there hadn't been a lot of that about, there was only one buyer of old sows that day. There was quite a crowd around the ring; perhaps they suspected there might be a laugh going, but only Eddie Johnstone, whose family must have killed more old pigs than any over many years, was buying for the pork pie trade. But while the auctioneer would take bids from the sparrows to make sure I wasn't done altogether, I was not happy with the bidding so I joined in and forced Eddie to a decent price. But when Eddie Johnstone worked out who his opposition was he was indignant – not outraged, but he had some theory that I should only be allowed one bid. Anyway I can't have gone high enough, for Eddie got the whole lot and I got the cheque I needed to buy feed to finish their progeny. There is no doubt that we both enjoyed the episode but only one of us enjoyed what happened next – and it wasn't me. The pig ring at Kittybrewster was circular and made of steel. There were vertical bars about six inches apart and a horizontal bar at the top and bottom held them together. Then there was another horizontal bar at about eighteen inches off the ground, a handy height for the buyers, or in this case the seller, to put his foot up on while his knee protruded through the bars. It was quite a common stance at the pig ring but the others who adopted it seem to have had knees that were thinner than mine. When my pigs were all sold Eddie Johnstone expected his adversary to give him a small luckpenny and then leave. Well the luckpenny was no problem, he had spent over £5,000 after all. But I couldn't leave. My knee was quite stuck. After the fun they all had at me I think all the butchers should have given me a luckpenny as I pulled and twisted, but my knee had clearly swelled up and I was stuck. I don't know how long it took or why, but eventually, when I had given up trying so hard, my knee came away. I am not the only one that has never forgotten my last sale of pigs at Kittybrewster.

So I still needed a commercial enterprise. An obvious one to try was the grazing of cattle. After all, half of Little Ardo had been rented by John Gyle, who bought cattle on the spring day and put them on Little Ardo's grass until they were fat. That required quite a bit of short-term cash, but no buildings would be needed. I was not so wet behind the ears as to think that I could just pitch up and buy the ones I fancied, put them in a park, shut the gate and wait till they were fat. I knew what they said about Patrick Wolrige-Gordon, our former MP who was famous for his support of something we distrusted called Moral Rearmament. They said that when Patrick lost his seat, and so could concentrate on farming, he too thought to fatten some stots. Along he went to Stirling mart with the cunning plan. He would buy them at less than the previous week's average price. When each stot came into the ring he had his calculator at the ready. He quickly dialled in its weight and up came his maximum price. If you are a farmer or a fish merchant, or perhaps even an antiques dealer, you'll understand this immediately; others may find it a little complicated. But believe me, it took the auctioneer about three cattle to learn what the young laird's maximum price was. He was then able to make sure that Patrick went home with the 100 worst cattle for sale that day and all at the market average price – far more than they were worth. There was no one else bidding of course, but a good auctioneer only needs one enthusiastic bidder, and one who looks at a calculator and then waves excitedly has had it.

Despite Patrick's warning, I can hold my hand up for a couple of blunders round the cattle rings, so you can see that there was much to learn. The first was when I went into the Belmont mart at Aberdeen and saw the dealer Willie Low selling a very mixed trio of heifers. They were quite stuck at £170, which didn't seem enough. I hadn't the sense to beware. I waved once and of course got them, as you always do when a dealer's cattle are stuck. That auctioneer was notorious as a bad speaker and in fact Willie's treasures cost me £270 each – I had paid £300 more than I thought. Then there was the time I went way out of my depth at Maud and bought the sweetest looking pen of chocolate Simmental cross heifers. Those were sold per head. I saw the weight, divided by three and decided that although they were dear they were so good as to be worth it. But there had been four in the pen, not three. They were good and they were one third dearer than that. Sandy Fowlie, of Adziel, comforted me, 'They'll grow awa fae that,' he said kindly. He was right, and by the end of the summer they were away back to Maud, where they made a handsome return. All of my blunders weren't disasters.

Cousin Stephen was good enough to let me into his plan for buying. He would position himself beside a group of big buyers like Charlie Coutts or Sandy Fowlie who bought the sort of cattle Stephen was after. When a nice penful came into the ring he would say in a loud enough voice to be heard but not to give offence, 'This would do fine for my low parkie', and then start bidding. His neighbours would politely leave him to it until he was either successful or gave up.

James Fowlie, the kind Sandy's brother, who himself bought thousands of store cattle each year, agreed that he would be my professor. He would teach me not to make a complete fool of myself, and eventually I might even make a little money at it. Actually he wouldn't do anything so simple as teach. He showed me how he kept cattle. I thought a farm with such a grand reputation as Auchrynie would have everything made of stainless steel, operated by remote control and connected to a computer which would give hourly printouts of the cattle's feed conversion rates and the prices of fat cattle at Maud and feeding barley at Buchan Agricultural Merchants. Well, it was nothing like that. The professor's guiding principles, so far as I could see, were that cattle should be bought right, sold right, housed as cheaply as possible and fed all they could eat. After that I was allowed to observe without guidance how the professor performed as a buyer at the mart. His only teaching aid was anecdotes, out of which the student had to extract the wisdom. The professor told me twice, so it was clearly an important lesson, about the Aberdeenshire farmer who had got a dealer to buy him some stots at Dingwall market. When the load of stots came home the buyer was delighted with the stock despite the fact that there were two very ordinary ones. He phoned his agent and after the usual pleasantries said, 'Aye, the maist o' that stots were fine but what were that two screws doing among them?'

'Oh,' said the dealer, 'it was getting late and I thought I had done well, so I just took those two to fill out the float.'

'Well anither time, jist draw in about till a heap o' steen gatherins and fill oot the float frae there.'

From that the student took that whereas you can't make much profit from a load of stones at least it is hard to lose money at them.

And he told me about the loon who found that it didn't take so long, because the cattle didn't drink so much if you watered your tied beasts with a dirty pail. I think that was about keeping water and feed offered to beasts as clean as you would like your own.

And there was the crofter who bought two calves from a dealer which sadly both died. The dealer was sent for and his comment was 'Man ye've had bad luck. But it could have been worse. I sold two calves to another crofter and HE died.' From that I think I was supposed to learn that whatever happens could have been worse,

197

and that it is better to make your own mistakes in the world of cattle dealing.

Professor Fowlie also taught me by *bons mots*, one of my favourites of which was: when the drink starts to interfere with your business, give up the business. I think that was meant to tell me that it is easier to cure ignorance to help your business than to reform your character. A strong man will find a way, but a weak one should give up and find something he can do. I took it to mean that, although James could see I knew nothing about marts and fattening cattle, I had the stuff that patience could yet make into a farmer.

He told stories and I listened and sometimes I learned. Even when I didn't learn I enjoyed the stories.

But that came later. It was as early as 1974, my first spring back at Little Ardo, that I went with Bertie Paton, my caber-tossing pal, to the great spring sale of weaned calves at Oban. The idea was to try a few stots on what was then the sixteen acres of the Single Hoose on the braes. Two stots to the acre would give me £1,600 less expense if I could buy them for £200 and sell them for £250 fat off the grass. Bertie, who was at this time still a ganger with Betts the Builders in Dundee, had rented a field off his father-in-law's family at Hayston, near Glamis, so we would each buy a load. Our day's work was a triumph and a disaster. Feeling well out of my depth, as Little Ardo had never bought store cattle at the mart and very seldom used the mart even for selling, and appalled really by the poor wee puddocks of calves on offer, I fell easy prey to Stott and Willox who, just as they had been when they saw the Chianina bull at the Paris Show, were at Oban because that was where the action was.

They invited me for a drink and they would show me how to play spunkie. It is the only time I have ever played the game but I remember how it is played. Each player gets three matches and holds out a hand with one or two or all of his spunks in it. So with three of us there could be anything from none to nine spunks in play. What the players had to do was guess what the total was. No doubt because of my ignorance I just won everything. The tactic which brought me success was to make the most improbable guess.

If I had one spunk in my hand I knew the most was seven and least one. Well I guessed there would be seven. Anyway that was profitable and very jolly. But there was another surprise due. In my absence Bertie had gone mad and bought a hundred stots of which my share was forty-two of the most stunted, hairiest, hungered, Hereford and Shorthorn cross Highland beasties I had ever seen. You just didn't get cattle like that on the east coast. The good news was that these little puddocks had averaged well below £100 each and the cheapest was only £69 for a little black puddle-jumper.

When these West Coasters got home to the lush grass on the Single Hoose park they grew a little but they put on weight like nothing else. The forty-two of them all went fat in quick time and before the end of the grazing season they had all made the eight hundredweights necessary to get the subsidy and had left over £100 a head. I had £4,200 profit off sixteen acres. This was clearly the way forward . . . and I didn't even give Bertie commission on his shrewd buying.

The next year I decided I would have 200 stots; if I looked after them well I could raise them £100 a head – £20,000 towards the overheads at Little Ardo. So next spring I was back at Oban, but this time I was on my way to the islands to buy badly fed cattle at source. Sadly the destination was Islay, where they knew a thing or two about looking after cattle. These calves had lived royally off waste from Islay's many distilleries and they were the opposite of skinny. They were polished. They cost much more than the Oban stots and when they got to Little Ardo's braes they said 'Wait a minute, where's my bran mash?' I found I did much better with the cattle bought in Aberdeenshire, which I had no difficulty in raising by my £100 a head. So Islay was a disappointment, but I did witness one event which will comfort me for as long as I can remember it. It was at the on-farm sale of a settler family called Epps who had an offering of more than 300 six-hundredweight calves which were looking well. And also looking well was the spread with which Mrs Epps greeted us at 9 o'clock on the morning of the sale. I was tucking into the dainties when Willie Low, the Aberdeenshire dealer, arrived looking none the better of the tremendous night he had had the previous evening at the Bridgend Hotel. Willie was a regular and a volume buyer and Mrs Epps was pleased to see him. She gave him a big hug and said, 'Now, Willie, will you have a dram or would you like a nice cuppie of tea?' 'Oh na,' said Willie, 'it's some early in the morning for tea.'

Those grazers did pay, but what the farmer of Little Ardo in the mid 1970s really needed wasn't just a suppy profit here and there, but some cashflow. It was all very well to take a team of bulls to Perth twice a year and come back with a good cheque. And while the export trade went on getting the occasional windfall, like when the two heifers sent to Canada on spec sold off the aeroplane for $10,000. But we needed something like the milk cheque, which my father used to get each month, to which he could look forward all month and which brought a bit of relief every month.

What I decided on was a beef production line producing a dozen fat cattle every two weeks. That would bring in enough to balance the books and to banish the apprehension of farming with a big overdraft and huge interest rates. The Aberdeen Beef and Calf co-op had developed a thriving market for bull calves from the dairies which would not be castrated as was almost universally practised, and would be fed intensively so that they would finish before they were a year old. I would buy thirteen calves per fortnight. Allowing for a death, that would give me a dozen bulls going off to the Waitrose supermarkets in England and a regular cheque each month. For that I would build a state-of-the-art fattening house. The bulls would be housed on slats so there would be no mucking-out to do, and no straw to be provided. At harvest, 200 tonnes of grain would be put into a fancy new grain tower. From there, at the throw of a switch, it would flow out into the bruiser and then to a mixer and then up an auger into a conveyor belt which would deliver it into the twelve hoppers. I used to fantasise about lying in the bed in the front room of the farmhouse, cocking an ear and hearing the augers starting up, and the bruiser, and saying conversationally to myself, 'Oh! There's Charlie feeding his nowt', while Charlie was snuggling down between the sheets and wondering if the postie would bring in the latest 'milk cheque'.

Sadly, I never did feed my bulls by throwing a switch by my bedside. There were just too many moving parts to my system and I never did get them all to work at once. In fact I never got the grain to go as far as where the hoppers were to go, so I never did get the hoppers into which the mixed grain would be piped down from the conveyor belt and which would switch themselves off when the hopper was full. Indeed, I made do with a bath for each pen, though where I was able to find twelve old baths I can't remember. The mixer was also a problem and I had to settle for the following system, of which Heath Robinson would have been proud: the grain came out of the tower and into a hopper from which an auger delivered it to the bruiser. From the hopper below the bruiser it went by two augers to a rickety wooden bagger on the floor of the grand new shed. There I filled it into bags until half the floor was covered in bagged bruised barley and the most delicious alcoholic smell would almost knock you out. Then twice a day the bags would be carried along to the baths and couped in. The scientific mixture of minerals and supplements was achieved by throwing in handfuls like bathsalts and giving them a stir with the hand like a lazy mother.

For all the faults my design contained, and all the moving parts that didn't move, the barley beef did fulfil its main objective of giving Little Ardo a 'milk cheque'. When the Aberdeen Barley Beef company were calculating the profitability they used to say 'Take the cost of the calf plus 10 per cent (a generous allowance for losses), add two tonnes of barley (which is more than they'll eat.) That gives you your costs. Take that from what you get from Buchan Meat and that's your profit.' That calculation always yielded a handsome profit which could be as much as £100 a head. That could be £25,000 a year . . . a fortune. I was going to be rich again every time I did that calculation.

Sadly, those were chickens counted but seldom hatched. The overdraft rose inexorably. The answer was to put in some more bull calves, do the calculation again and show it to the banker who had never seen a cashflow. It was always set out neatly, showing how in a year's time the cash would be flowing at him so hard he

201

would need to brace himself against his big desk. Sadly, cash projections are only done forward. They are only as good as the information you put in and the assumptions you make. If you put in rubbish what you get out is rubbish too. No attempt was made to look at the actuals . . . to see what actually happened. What was the banker thinking about? What was I thinking about? Were we thinking at all?

It is too late now, but I have come up with some clues as to what went wrong. The 10 per cent allowed for losses would have worked for the dairymen who reared only their own calves for their barley beef units. But I had several good dairy suppliers and so increased the pool of disease from which my calves might die. And my need for calves was so great that I would take them from anywhere. I couldn't see a penful of Friesian bulls at the mart without bringing them, and everything they had caught at the mart, home to Little Ardo. Then I got some wonderful-sounding bargains from calf dealers who had bought them in England, peddled them around several marts and were now ready to offer me a good deal, having failed to sell them elsewhere. Those calves offered me every disease known to man. And my wonderful concept, which so pleased the theoretical economist who still lurked within me, meant that as the cattle circulated clockwise round the shed in a continuous flow towards fulfilling my quota of fat cattle going to Buchan Meat, so the diseases flowed continuously anti-clockwise round the shed. I gradually came to see the point in batching cattle. That way you can give your sheds a rest between batches, and a spring clean, to break the chain on which disease passes from old to young.

All the same, though it was never calculated with precision, I think I did quite well on deaths at the calf stage, and the 10 per cent allowance was more than enough. But no one had told me to allow for what turned out to be the real scourge. Cattle housed close together will often tramp on the carelessly laid tail of one of its fellows. If they are bedded on straw that matters not at all. But on slats it can lead to bleeding and infection. Again, with heifers and steers, though it may be painful, no great harm is done. But for reasons that no one could explain to me, bulls with trampled tails will get infection into their spines and they will die – not all of them, but we did seem to get an epidemic. Now you can see at once that this was a serious blot on the cashflow – much worse than anything we got from the calf house at the mart. A calf lost is perhaps £60 down the drain. But a bull as it approaches its market weight would have been £500. It didn't take many of those to spoil the balance sheet – and there were many.

And how does a cashflow deal with what happened to the moist grain store? We had filled it with barley with a moisture content of 25 per cent, which we thought the ideal for digestibility. The last thing to be done in such a job is to shut the tower and thereby seal the grain in. The loon was sent to climb the thirty-two feet up to the top of the tower 'and be sure that the seal goes right round and it's screwed down tight'. When he came down I asked him if he'd made sure it was tight and was reassured. It was probably the worst single piece of bad farming among many that I did during my tenancy of the little farm on the hill. No job as important as that should be left to the loon. Jimmy Low would never have trusted that job to anyone. He would have done it himself. All was well for a while and then we started to get bits of rubber coming through in the bruising. Then several chunks of the seal appeared. It is fairly clear that the loon, while trying carefully to fix the seal, had dropped it in amongst the barley. It is a sentence of death to go in after such a seal but I did wish he had told me at the time. We had a spare seal and could easily have put it on, but now it was too late. The barley began to change colour from golden to an insipid grey and the gorgeous smell of moist grain changed to a fooshtie choking stink. It ran for a few more days after we put on the sweep auger which runs round and round the inside of the tower and can get hold of grain that is too wet to run. But eventually it stopped running even with the sweep. I opened the hatch at the bottom and looked in. It was quite a sight. The barley had bridged, reaching up to a peak of perhaps ten feet in the centre and making a perfect Roman arch to where the sweep auger had cleared it right round the edge. I wanted to get in and poke up at it with a stick but that would have been very dangerous. I tried firing the twelve-bore up at it and that brought down some but not enough. In the end we had to break into the tower at a height of about twelve feet and put in a temporary auger and poke it with sticks. Thank goodness there were only about 100 tonnes of barley in the tower and eventually we did get it all out. The bulls ate it but the feed value was very low.

I never thought, when I was contemplating returning to the farm, that the day would come when I would be running the fifteen miles up to Turriff to Buchan Meat to collect my cheque for the latest load of barley beefers. It wasn't that I had run through my credit and couldn't get any more. The bankers still hadn't twigged that my cashflow projections were rubbish. No, the dash to Turriff was justified by the fact that the cheques took two days by post and always arrived on Saturdays. Going for my cheque saved four days' interest on the money. If you are paying 20 per cent at the bank a cheque for £7,000, which was about the usual amount, was costing over a £4 a day – and £16, in my reduced state, was well worth the run to Turriff.

Sadly, I was too late to ask Old Maitland Mackie if it was wise to go into barley beef. He had been a pioneer of barley beef twenty-two years earlier. He would have got an automatic system to work all right if that was good idea but he also knew the great flaw. There is a very interesting item in the weekly letter, a copy of which he sent for thirty-five years to each of his family. There he wrote of his barley beef, 'I sent a roast to (three well-known beef men) for their opinion. I think I know what it will be. The beef is tender but tasteless.' I eventually discovered this myself as a result of the thrifty solution to the tramped tails problem. We decided to put a barley beefer into the deep freeze and eat it ourselves. It was, as I had been telling people for years, tender and had no fat on it, but it tasted like a Unilever product. It melted in the mouth but left the taste of soap.

CHAPTER EIGHTEEN

Producer at the BBC

1981–86

I was discovering that farming was not as easy as the calm demeanour of many farmers makes it look, but all was not going badly for the latest tenant of Little Ardo. I was developing a broadcasting career.

I am often asked how I got into broadcasting. Well, the answer is quite simple, but I don't think it is much help if people are really asking for guidance as to how to go about getting a job at the Beeb. My case went like this . . .

I won the Supreme Simmental Championship at the Perth Bull sales and was interviewed by Allan Wright about my success. I did not say 'I'm over the moon.' I didn't even say, 'We knew we had a good bull but I never really expected this.' I didn't list all the others who had contributed to the bull's success and I most certainly didn't say that some of the credit would have to go to Sandy at the Meat and Livestock Commissions' station at Aberdeen because he had fed the bull for several of his most important months.

What I did say was that even now that we had the Charolais and Simmental breeds established in Scotland, the continental invasion wasn't complete. I had a scheme to use the Marchigiana, an enormous white thing from Italy, related to the Chianina but not quite so tall and much beefier, to breed calves for my barley beef unit. These beasts would be lent to dairymen for use on their heifers. That would work well as the calves would be small and slim and I would give a good price for the calves. Being lean and light-boned, the bulls would yield great returns to my barley beef unit. (The plan was one of my little disasters. The calves were, like their parents, huge, and having given birth to these monsters the farmers were reluctant to take the decent price I had offered when the live calves were clearly worth a wee fortune. But that's not the point.) It may have been a hare-brained scheme but it was interesting, and Wright thought his guest spoke well. So he asked if I would come to the next Perth sales and give a wee comment on all the main events. That went all right, so he asked me to do a weekly five minutes on the newly opened Radio Aberdeen on whatever grabbed me from the farming scene.

That went well for the farmer of Little Ardo. In the late '70s everybody was telling farmers they needed to diversify so this was my diversification. The cheques from the BBC came in handy. I used to get about £25 for my piece on Radio Aberdeen. But then Wright played it again on the *Scottish Farming News* on Radio Scotland and I got paid again. Then once a month Wright did Radio Four's *Farming Today* programme and he used it again. It was always picked up by Radio Scotland's *Best of Scottish* and sometimes it was on Radio Four's *Pick of the Week* programme. On a good week I was being paid five times for the same five minutes' work. It certainly made the run into Aberdeen an acceptable sacrifice and it seemed to beat farming.

Anyway, the stock of stories, often just yarns John MacIntosh or Professor Fowlie had told me at the mart, was lasting well and the flutter of little cheques coming through the letterbox made me think I could do more to follow in the footsteps of John Yull the auctioneer and John Allan the writer to earn income off the place, though I had no real idea as to what I might do. So I told Allan Wright that if he had any ideas I'd like to do more of this broadcasting. It was a total surprise to me when he said that he'd give me a full-time job on the staff of the BBC.

What had persuaded Wright that I should make this move was another of his schemes for Radio Aberdeen. He had had the idea of cutting John R. Allan's classic, *Farmer's Boy*, down into manageable chunks so that I, the author's son, could read them and he would broadcast them. He was convinced that that had gone well enough to justify our trying to do more broadcasting together.

This called for a reorganisation of my farming enterprise. I had already had a menagerie sale at the mart in Aberdeen and got rid of the remains of my Chianina, Marchigiana, Romagnola and Gelbvieh at commercial prices and had already sold the Charolais at Perth at up to £3,000 a head. The grazing cattle were not a problem. I would sell the grazers as they went fat and buy none next spring. I would stop buying Friesian calves and let the 300 barley beefers gradually run themselves through the system and into the overdraft. I still believed in the Simmental breed, though it was becoming more and more apparent that only those who could afford the best show cattlemen, for which they would need the best equipment, the best house and the best wages, would succeed, and I could see that wasn't going to be me. Reluctant to give up, I decided to sell the lot but to retain a stake by a sneaky move. I would put my Turriff Show Champion, my Highland Show Champion and Gertrude, my best breeding cow, through my transplant unit. The eggs would be transplanted into black Hereford heifers and I'd be back in business with a Simmental herd all bred from my three best cows, in a matter of three years. The break from sales and shows would let me have a go at the BBC.

My CV was hardly typical of a staffer at the BBC. Most of those who broadcast, your Alan Hansens, your Russell Brands and even good broadcasters like David Attenborough, are not on the staff of the BBC. They are employed by the staff to present particular programmes. It would have surprised no one if they had taken me on as a commentator, a summariser or one to make a fool of himself from time to time or even often. But only an original thinker like Allan Wright would have thought of me as the man to organise, not just whole programmes, but make sure that studios were booked and contributors paid and Scottish agriculture covered in a satisfactory manner, all the year round.

So people were watching. Wright didn't want his new appointee coming in wearing sharny boots and smelling of silage. 'You can't turn up at the BBC looking like you'd had a rush to get in from the byre to get ready for a day at the mart.' He needn't have worried. I have never been a snappy dresser but I had survived thirteen years on the staffs of various universities and never once appeared without a tie and shoes, the leather of which, if it didn't exactly shine, could at least be seen. I knew I would be expected to conform to the dress code and I was ready to meet Allan at least halfway.

But I didn't realise that this smartness idea would apply also to the new Farming Producer's car. I had long since decided that my farming could not support a big car that would bring admiring looks from my peers and had settled for something that would go. This was a Vauxhall Viva of long vintage taken by Lawrence of Kemnay in a part-exchange and not touched by Lawrence in any way. They just wanted shot of it and the price was £100 on a buyer-beware basis. My new boss didn't like my new car. I don't know if it would have taken on the sort of polish Allan Wright would think appropriate, for I never did try. But it was when the old car blank refused to start one morning, and Fiona was off to do battle with the kids of Turriff in her smart Peugeot so I couldn't borrow that, that I really got in trouble. I was a hard man to stop in those days and with both cars out of my use I sought desperate measures. Allan Wright was not amused when his fellow producer turned up at Beechgrove House in the old cattle float which had by this time lost much of the attractive sky-blue paint with which I had covered it to cut a dash at the shows. It had been Maitlands' (the upmarket furniture store in Turriff) old delivery van. Ewen Booth, the farmer of Downiehills, had bought it for his daughter Susan who, as well as being a very pretty girl, was a classy horsewoman who did well enough in showjumping to deserve better. It was an Albion, the diesel engine was in great nick and it had partitions which could provide separate stabling for four horses, as well as a gas ring and room for a bed that could be improvised above the cab. As Ewen said when he was selling it to me, 'It's got a teapot an aathing.'

The old Albion became somewhat notorious because of the way our pedigree cattleman Gordon Paterson used it at the Highland

Show, the Royal Show and the Bull sales. At first it was known as the 'Albion Hotel', but it later became 'The Gordon Arms' as the legends of the parties held there grew. There was always room for one more at that inn. One of the disreputable stories about the Gordon Arms was about the innocent who, like everyone else, got far too much drink, and was getting gracious with a lassie who was helping look after some Herefords. One thing developed into another until the light faded and our friend found himself in a fond embrace. And our friend wasn't perhaps as much of an innocent as we had thought, for when the lassie's fondness grew into passion he thought this was surely a sign that he should try his hand. That was a disappointment and a revelation. He found that someone else's hand had beaten him to it.

At any rate, Allan Wright of the BBC was not impressed when his new producer turned up in the Gordon Arms to do his shift as the voice of farming in Scotland. Parking was always scarce at Beechgrove, which had been built as a private house but was now a very busy little Broadcasting House. There really wasn't enough room round the back so I had to park right in front of Beechgrove House. And I couldn't park as indicated by the lines for the guidance of visitors, because with the van being so long the front of it would have been almost in at the front door. No, it had to be parked longways, taking up three parking spaces normally reserved for visitors and the news teams. Wright told me, in a way which did not lend itself to any misunderstanding, that that heap had to be removed and that it was not to come to the BBC again.

At the BBC I tried to get as much as possible of what they called actuality into what I did. I think the first of that was when Wright and I had the unorthodox scheme to explain the grass lets and store cattle markets and fatstock rings by getting out there and doing it. We went to a grass let and rented a field. Then we went to store sales, where I bought half a dozen cattle, and he bought half a dozen cattle and John Paton the Aberdeen dealer bought six cattle. We then had a competition to see who would make the most money. Each week we went out to our field with the tape recorder and had learned discussions about how the market was going and how much better my cattle were looking than Allan's. We made some money for ourselves though we didn't give John Paton any because we were confident he had made money out of his before they arrived at the field. There are no detailed records of who won but I am sticking to my version. Throughout my short broadcasting carreer I carried on getting out and about as much as possible. If I was at the Lairg lamb sales I did all my interviews outside, where you could clearly hear in the background the 30,000 North Country Cheviots and 3,000 north country voices. When I went to visit a turkey farm near Edinburgh to talk to the farmer about the prospect for the Christmas trade of 1984, I did the interview in the pen with 200 turkeys which were almost ready for the market. The turkeys were interested but they didn't make a great noise, just scraping about among their deep litter as they watched us. But if you spoke loudly they would all reply at once, so I got a wonderful recording. When I went into the pen I said 'Hello, boys' in a loud voice and they replied with a ripple of 'gobble, gobble, gobble, gobble', then we got on with the interview to a contented background of rustling, scraping and the occasional flutter. When the farmer told me some of these birds would make a pound a pound for Christmas dinner, I turned to the flock and said, 'What do you think of that boys?' 'Gobble, gobble, gobble, gobble' rolled round the tin shed in what sounded like indignation. And when I finished I thanked the farmer and then turned to the flock and said, 'Thanks, boys, Merry Christmas,' and they replied as one, 'Gobble, gobble, gobble, gobble.' I was pleased with that. I had filled perhaps five minutes of my daily programme in a satisfying manner. But I could have dealt with the information involved as a news item of eight seconds, by reading out, 'And finally, a Midlothian farmer has told *Scottish Farming News* that his turkeys will fetch up to a pound a pound at the farm gate for this year's Christmas market.'

When covering the Paris Show I recorded a street barrel organ played by a very fed up-looking monkey. That music could only have come from France, so it added atmosphere. That was what you were told to do and I liked doing it, but my successors who do it now drive me off my head. The background music, or whatever, sounds more interesting than what you are supposed to be listening to, but you can't quite hear it. Or else the background is so loud you can't quite hear what you are supposed to hear. I really shouldn't throw things at the telly, for I have done it all myself. Another of my attempts to bring actuality into the Farming News was aimed at getting over the tediousness of the long lists of class winners and runners-up that had to be read out at the big shows, not getting anybody's placing wrong, never mind mispronouncing their farm name, even if it was in Gaelic. It was deadly dull and really not of much interest to anyone except the few who had won prizes, and their few friends. I decided what was needed was more actuality. When the supreme championship was being decided I would get into the ring and over all the background noises (mooing and people saying 'Fit the hell's Charlie Allan daein in there?') explain how each of the competitors had got to the final. It was still pretty boring but at least I was trying and, just once, I did succeed in raising interest levels – by more than I meant to. In cattle judging the only remotely dramatic moment is when the judge has chosen his champion, usually after a long wait of simulated uncertainty. He will then approach his champion from the rear and clap it on the rump, signifying its triumph. Sometimes the judge will approach as though to clap one but divert at the last moment and clap the other – for effect.

So there's Methlick's man at the BBC in the ring. In hushed tones he says 'Now the judge, Mr Expert from at least 100 miles away, has his champions before him and as they parade around to a packed gallery I have time to tell you how these two beauties made it to this stage. The senior bull is Roxburgh Roger from Burgh farm. He won the senior bull class nearly three hours ago . . .' And with impeccable timing he has given all the early rounds when Mr Expert is ready to make his decision. The excited commentator never falters: 'And now Mr Expert is going round the back for one last look at the two of them, the massive champion bull and the more feminine but still enormous senior cow are a picture. Which will he choose? He likes the cow. I think he likes the cow. Mr

Expert is approaching the beautifully formed hindquarters of the cow – will he give her the clap?' whispered the excited commentator. Well, he did slap her on the rump. It was a very well-received broadcast and my superiors at the BBC thought it was wonderful too, but then, as usual, though they must have known what 'the clap' was, they hadn't a clue what was going on.

There were many exciting episodes of my time at the BBC, not all of which reflected well upon the then farmer of Little Ardo, including the time I did *Scottish Farming News* from Selkirk. I sat nervously in the little studio waiting the call from Radio Scotland in Glasgow. Eventually it came. 'And now it's time for *Scottish Farming News* with Charlie Allan.'

'Hello, again,' began the brave hero. 'Charlie Allan here with all the latest news, views, prices and trends for Scottish farmers, which comes today from Selkirk.' Except that I had forgotten to book the lines to send my programme up to Glasgow. I was blethering away in Selkirk but the poor listener heard nothing. He was connected to the empty studio in Aberdeen from which the programme usually came. Fortunately a quick-thinking engineer in Glasgow managed to patch me in after half a minute or so by ignoring all his training not to do anything unless he had it in writing and in triplicate. The farmer-turned-broadcaster was so much at the mercy of the goodwill of people who knew how broadcasting actually worked.

It was a good idea to be friendly with the engineers, though many producers took a strange satisfaction out of being unpleasant or at best ignoring their engineers. I always made a point of the elementary, but quite unusual, good manners of introducing my technical staff to the guests who came in to record their wise views on the programme, and I am sure that is why they took pity on me when things went wrong.

It happened often enough, but we were never so indebted to our engineers as in London in 1982. We were there to cover the Royal Smithfield Show. It was my first time at the fatstock show at Earl's Court in London, and Allan Wright was in charge. We recorded interviews with the early show winners, with the machinery

212

companies who were launching new products and with the politicians who were there to plug their views and the Farmers' Union members who were there to criticise them. We had even taken the livestock market reports from Maud and from Stirling. Allan Wright wrote out his script word for word so that all he would have to do was appear in the mobile studio and read the script at 6.15 the next morning while the engineers played in the tapes of the interviews we'd done. Then it was off to the town for Wright to show his assistant how to enjoy himself in London. That went well. There was a great dinner with all our fellow agricultural journalists and a Simmental Cattle Society party, and it finished in our hotel where our excellent secretary (and market newsreader) Alison Cruickshank discovered the 'Captain Bar'. Suffice it to say that in the small hours, which were by this time not that small, I had to tear Alison away from the Captain Bar, which she had mistaken for a one-armed bandit. She had discovered that if you dialled in the right numbers you could get any drink you wanted. 'Twenty-five – gin and bitter lemon. Seventeen – bacardi and coke, hee, hee, hee. Och, Charlie! Ye canna stop me when I'm winnin.'

Dawn broke slowly on me. There was no rush. Wright was on duty and would present the programme. I would have a shower, a leisurely breakfast and saunter along to Earl's Court about half past nine. It dawned only gradually that Allan Wright was still in the other twin bed. I woke him and in great alarm we tried to sober up and focus on how we could get across London and into our studio. It was impossible. The two producers contemplated their P45s. Then in wild despair we decided to put on the radio to see what music they had put on to fill our twenty-minute slot. It took a minute to get Radio Scotland but when we did we were astonished. Up came this very Cockney voice reading out very deliberately '. . . so I sent our reporter Charlie Allan down to ask the winner what he thought of his success'. And up came a very Aberdeenshire voice saying that he was 'fair tricket'. Then up came Charlie Allan's voice asking him if he had been surprised by his success. 'Weel, I couldna very weel say that. If I didna expect tae get a ticket, why the hell would I bring a beast aa the wey tae

London?' Wright and I looked at one another, standing there in our pyjamas, and wondered what on earth had happened.

Of all the people at Smithfield Show, we were just about the last to know. There were two engineers (called studio managers) on duty. As they are paid by the hour the engineers are always in good time. Five minutes would do all their checks and they would expect the producer and his presenter – in our case they were both the same people, as Wright and I were producer/presenters – to appear at least half an hour before yoking time. But there had been no sign with half an hour to go, and not at twenty minutes before the off. With ten minutes to go, one of the engineers, Julian, took a wander into the studio where he found the script neatly placed where it would be needed beside the microphone. He read a bit so that Steve, his mate, could check that the microphones were all in order. Then they decided that if we didn't turn up they would just do the programme without us. We didn't, so Steve played the tapes and Julian read the introductions. He had no previous broadcasting experience. He broke all the rules, as did the engineer who was not supposed to drive this important station single-handed. Best of all, they broke the unwritten law that says, 'Unless it is in your job specification you will do nothing to dig those fancy bastards on the production staff out of any holes of their own making.' Both could have got the sack. They didn't, but I really think Wright and I would have got it if it hadn't been for those two being such troopers. In fact the only official word we heard about it was from the listeners' panel who met every three months for free drink and a chance to criticise what we were doing. Someone at the next meeting did say to Allan Wright, 'I didn't think much of that chap you tried at Smithfield.' I still don't know if she knew all along and was merely pulling our legs.

It was all a big change from the farming. I met a lot of interesting and different people. Like the two young trainees from Sweden who came to Aberdeen to learn from our studio managers how to engineer programmes. They sat in all day watching what two young studio managers, Stuart Bruce and Doug Maskew, were doing. They got on with their jobs and didn't think much about it until they finished their shift, which meant the girls too were finished for the day. As no more than a polite afterthought the boys said to the students, 'We're just going over to the club for a drink. Would you like to come?'

'Oh yah,' said one of the flaxen-haired beauties, 'but we are not sleeping with you.' I don't know what she made of the studio manager's reply, which contained just a hint of outrage worthy of an older man, 'You might wait till you're fuckin asked,' he said. This older man wouldn't have known what to say.

For several months the farmer of Little Ardo had a fine easy life at the BBC. Allan Wright liked doing as much of the broadcasting as possible and I got all I needed to help reduce the overdraft on the farm. I suspected it couldn't last, and it didn't. Wright got promoted and moved to London to *Farming Today*, Radio Four's farming programme. I was left to do the work which had hardly been enough for the two of us. It was easily enough for one. I not only had a programme to do every day from Aberdeen, but I also had to do *Farming Today* on Radio Four on a Monday to give Wright, who had left me in this state, a day off. On top of that Wright had invented an extra programme called *A Story, a Song and a Poem* in which he read a poem by Borders farmer Tim Douglas, Jean Redpath sang a folksong and I made up and read a story. It was a popular programme and with Wright keen to hog the airtime, I had had plenty of time to get on with writing wee stories. But with Wright gone the inexperienced broadcaster had no spare time and no hope of a day off for lack of anyone who could fill in for him. I had no one to let me away to my daughter Sarah's graduation in Edinburgh, and when my Aunt Lindsay died young I went to the funeral but couldn't stay for the wake as I had to rush back to Aberdeen to present my programme. There really was no one else. After another year in the job I could have fixed things up to go to an Edinburgh studio to do the programme and still made it to the graduation, but at that time I was struggling to do the job at all, never mind to do the work of two. After six months which seemed like a lot more, Jeanne Gavin joined me as a colleague and the job became a dawdle again, but it had been a nervous time.

The BBC was pretty well a full-time job so Little Ardo got little attention. The eighteenth-century mansion was still a great place to stay, but my mother's cousin, Stephen Mackie, grew crops on half the farm. Stephen told me that Little Ardo was 'a willing place' which I took to mean, 'This place could produce a lot more crop if it were properly farmed and that I am about to do.' I think he did. I had a good rent and I certainly enjoyed going off to work in Aberdeen through my fields all growing good crops, even if someone else was getting any surplus. As I had never managed much surplus, it was no skin off my nose. Stephen grew the barley and wheat and left the straw for my cattle which inhabited the steadings in winter and the parks in summer. The only staff was one man, who paid rent for one of the cottages and worked part-time. 'Basher' Burnett is a special character who deserves a book to himself but won't get one. He is a gentle family man with a reputation, earned in his younger days, for clearing bars if he didn't like the company, and for whom no task was too difficult. He made a good job of keeping the cattle alive, but there was no fancy feeding or showing of cattle.

CHAPTER NINETEEN

The Overdraft Wins

1981

I got sair back and arthritis and a stiff neck frae the ploo,
And farmer's lung fae feedin hay and breathin in the stew.
I was aboot as much use as a brush without a shaft.
I was at a loss. My albatross – colossal overdraft.

My father worried all his life about two things. He had a horror of
being consumed by fire, which I thought maybe had something to
do with the colourful way he had been taught about heaven and
hell, and he worried about his overdraft, which was certainly a
reaction to the way his adolescence had been dominated by his and
his widowed granny's poverty when the great Allan dynasty at
Bodachra came to an end. It would indeed have taken a jealous god
to consign John Allan to the flames, and that overdraft was never
more than £15 an acre – but it worried him. I, on the other hand,
having been brought up a relatively big farmer's son and having
rubbed shoulders with Maitland Mackie when he was building up
an empire of 3,000 acres, all on overdraft, was unimpressed by debt
as long as I could show that there was something to put against it.
My coolness in the face of mounting liabilities may not have been
an unmitigated blessing, for it helped my debt on Little Ardo's 252
acres to spiral upwards. In round figures, counting my main
overdraft at the Bank of Scotland, a smaller overdraft at the
Clydesdale and 200 cattle bought from the mart on a seasonal
basis, it peaked at £1,600 an acre. When an acre was worth perhaps

£500, that was a lot. When interest rates were over 20 per cent that was £80,000 a year, or to put it in more familiar terms, it was equivalent to a rent of £320 an acre. Only a very good farmer indeed could have paid such a rent and knew I wasn't that.

And yet when I went to the banker I was always able to show that if I sold the 200 grazers, and the well-known Ardo herd of Simmentals and the 300 barley beefers, the whole position, or most of it, would unwind and my debts would be seen to be manageable after all.

By 1981 I had already done a good deal of unwinding. The Charolais had been sold off when I decided I wasn't fit to compete in that breed. The exotics that I judged had nothing to offer if the Americans were no longer interested, were sold. That got rid of the Gelbvieh, Chianina, Romagnola and Marchigiana. The Blonde D'Aquitaines were a very good trade. One died and the other was infertile and they were insured for £5,000 each. The pigs were away, but they just made the money at which they were in the books and that was nothing. At least £100,000 of overdraft, the cattle bought as grazers would go fat off the grass, and then be sent to the slaughterhouse and need not be replaced next year.

The job with the BBC had forced my hand. My excellent pedigree cattleman and grieve, Gordon Paterson, had really sacked himself by refusing to heed my warnings about enjoying himself too much at the bull sales and at shows. And Willie Adie had joined the exodus of the old guard of Jimmy Low and John Allan, so while I was capable of a lot of work, I couldn't keep on a herd of pedigree Simmentals and make a job of them as well as learning a whole new job at the BBC. But I had a plan – a plan so clever it seemed doomed to ignominious failure. I would sell the Ardo Simmentals in such a way that they would all be home again in three years.

I wanted to sell my herd in a 'complete dispersal'. That is the way to get the best prices, because the buyers know everything is for sale, not just the ones you don't want. But I was proud of the Ardo Simmentals and wanted to think that one day I would be back at Perth showing a champion and topping the sale. So I took the best three cows, and using the facilities offered by my mobile transplant unit I soon had thirty-six black Hereford heifers pregnant with those three's calves. Then, at a great sale in Edinburgh to coincide with the Highland Show of 1981, I sold all but the transplanted heifers. That grossed £40,000. But it was the most incomplete herd dispersal because I got thirty-five live calves all pure from my transplants and they formed the basis of my new pedigree herd. I really enjoyed that. It reminded me of the way my father and his grandfather had made money seventy years earlier out of their homing pigeons. Every time a family including a little boy came, my father would present the child with a pair of his beautiful white doves. This magnificent gift was always rewarded with a shilling, or even half a crown, but as sure as death, they had always returned by the next morning ready for the next charitable donation.

As my barley beefers went fat I stopped buying bull calves to replace them and barley to feed them. That worked wonders for the cashflow. So by the end of the year of 1982, the stock was down to sixty head of cattle and no pigs, and the overdraft was down by three quarters, to no more than £100,000 – a mere bagatelle.

My job at the BBC could almost pay the interest at the bank. With Fiona now a senior lecturer at the College of Commerce we were solvent for the first time since we had left Glasgow. I had never noticed that my overdraft was a worry, but getting it to manageable levels felt so different that I must indeed have been worrying.

CHAPTER TWENTY

Househusband in Africa

1986–89

We had always talked about a job in the sun. When we met, Fiona had already applied for a job in Uganda to lecture at the famous Makerere University. I could have gone to the University of the West Indies to lecture but they said that would be a bad career move. I had been invited to apply for a job as the economist at the Treasury in Jersey but at 25, and having won none of those major gongs yet, I didn't want to give up on the Highland Games. And later on there were the four children. They were always either settled in school or unsettled in their work and it seemed we should stick around. But by the mid '80s the farming, which still wasn't paying, at least was not overhung by an impossible overdraft, and my track record at the BBC and the universities would make it possible for me to work anywhere that English was the main language. The children were all busy doing things we really couldn't influence or could influence just as well by telephone and mail. Sarah was married and practising her trade as a chemist in Aberdeen, John was on an oilrig, Jay was a beach bum on the Costa Fortune, and Susie was at Emory University on a golf scholarship.

Fiona had a very good job in computers, a set of skills she could also take with her anywhere. With the aid of a roaring fire and a bottle of malt whisky we made a pact one night. One of the problems had always been that while we each had had chances of a job in the sun, it was a bigger problem getting a chance where both of us could work and at a moment which suited both of us. The next time one of us got the chance to work in the sun, we agreed, the other one would give up their job, go along and see what could be picked up when we got there. Knowing that Fiona would favour Africa, not fancying any of the Arab countries and afraid of the troubles all over black Africa, I did stipulate that in Africa I would only consider Kenya.

I immediately forgot about our convivial evening, and our pact, but not many days later we were having breakfast together in the kitchen at Little Ardo and I wasn't listening as well as I should, when I heard the voice at the other side of the paper say, '. . . but it's only for two years'. I put down the paper. I knew I should have been listening properly. Fiona had applied for a job in Kenya installing and maintaining computer systems for marking school exams. She would get a big salary, a house and a total change. What could the farmer of the small uneconomic farm on the Buchan hillside do? I gave in my notice at the BBC. Stephen Mackie made me a very generous offer to take over the whole place on a seasonal basis, which I accepted. Little Ardo would still be in the family after all. The thirty-six transplanted Simmental embryos had produced thirty-five pure cattle, including seven bulls for my last Perth bull sale. I was able to put my Simmental herd on the market for a second 'complete dispersal' in three years. I got an oil man to rent the old farmhouse, fully furnished, including Gladys (who had been our housekeeper since 1976, when Fiona was driven out to work by our overdraft). She would stay on to ensure that the tenants didn't destroy the old house completely. Everything that could be stolen when we were away was sold in a grand roup ('the farmer's life laid out in lines'), the first at Little Ardo since Dod Yull left for Australia in 1911. The dispersal of all those transplanted cattle, this time at Macdonald Fraser's at Perth to coincide with the October bull sales made £38,000, the embarrassingly poor roup of my pathetic machinery some £7,000, and with the few remains of commercial cattle all that came to a further £50,000 reduction in the Little Ardo overdraft. It took about six months, but after that I was off to the sun to join Fiona and Little Ardo was relegated to absentee landlordism.

Africa was a culture shock. Suburban Nairobi was very different from rural Aberdeenshire, or even suburban Dundee. But the shock was not what my friends had expected for me. I had been told that as a spouse of a foreign aid worker I would not be allowed to work. Everyone predicted I would go off my head with nothing to do, but nothing could have been farther from the truth. I took to the life of leisure as to the big hoose born. When it came to doing nothing I found that I was a natural. I joined the Nairobi Club, which had been set up in 1907 to cater for the leisure and sporting needs of the top men in the British Colonial Service. It has the most magnificent cricket ground, bowling green, swimming pool, billiard room and squash courts, and the tennis section, with its eight courts and six hundred-seat stadium, is where they hold the Kenya open tennis tournament. At the last count there were eight bars and places to eat, except that when there are functions they will pitch a tented village anywhere on the club's twenty acres, with a full range of indulgence. The club now has a business centre but even in 1986 it had a well-stocked library and a full range of London newspapers. The farmer from Buchan found it very easy to wave goodbye to the wife as she left to brave the early morning traffic to the National Examinations Council offices in the city centre, make his way down to the club and sleep away the morning under a newspaper. *The Times* looked well, though I seemed to get a better rest under the rosy glow of the African sun filtered through the *Financial Times*. I didn't make as many friends as I might have in the club – my socialist upbringing was not the best preparation for dealing with my fellow members, mostly fading colonialists and Africans on the make – but there were enough to ensure a game of billiards and a jolly lunch, followed by another session under the *Financial Times* before I staggered off to the house to welcome Fiona home and tell her how I had been run off my feet all day.

It was the life of Reilly, but I did have my duties. I was a house-husband, peeling-nose style. I had staff to look after. The garden was extensive and ran to a gardener called Joffrey. Every morning I had to give Joffrey his orders, which included climbing up the avocado tree and getting two for our supper, or if the season was well advanced and the avocadoes grown to their full weight of more than two pounds, one for us to share. The fruit were delicious and very different from the poor wee ones bred down and sprayed so that they taste of antiseptic that we get in Tesco, and our favourite tree had an estimated potential output of a tonne of fruit a year. That the tree never reached its potential was because the fruit were delicious from the time that they were the size of large chestnuts up to when they were as big as coconuts. I sent Joffrey up the tree every day because the fruit was so good, but also because I liked to watch him do it. Like our arboreal ancestors, he climbed effortlessly the sheer trunk of the tree, which had no branches below about ten feet.

Brought up in wartime and post-war Britain, and ashamed to waste food in case it would impact unfavourably on the starving millions of India, I was uncomfortable at the abundance of avocadoes. Surely the least I could do, now that I was so much nearer the starving millions, was to give away as many avocadoes, which our African neighbours called 'poor man's butter', as they would take. We put about half a hundredweight outside the garden gate with a sign saying 'karibuni' or welcome. When they were rotten we took them in again. I realised then that though there was a lot of poverty and even some starvation in parts of Kenya, it was not a simple matter of overfed Europeans eating up the crusts of their toast.

Joffrey kept the garden and washed the car, inside and out, every time it came to a halt. But that still left the housework. There was no vacuum cleaner. No polisher for the parquet flooring which extended throughout the bungalow. No dishwasher. No washing machine. I had to look after all those chores with just one domestic appliance – and he was Alfie, a delightful man of the Luo tribe. Alphonse Oile Okech was a farmer from the shores of Lake Victoria, where he grew food for his family and kept cows for milk and as a rich cash crop. For the previous twenty years Alfie had left the wife to look after the shamba (farm) while he had worked in Nairobi, three hundred miles from home, for eleven months a year, to earn money for the education of his five children. When he had first left the lakeside he had had five cows and their calves. But as the family grew he had had to sell one and then another to keep up with the school fees. By the time he came to us he was down to one cow – but soon she had a heifer calf, and what a great day that was.

Alfie worked from sun-up, which was always about a quarter past six, until he had washed the supper dishes, which would usually be before eight. He gave us breakfast in the morning and made the best fish curry in Nairobi. And when he wasn't cooking he was cleaning, and when he wasn't cleaning he was polishing, and all at great speed and with this complete set of shining teeth that he kept in a smile that lit up the room. That wasn't quite right. The smile which did light up the room was not complete. The Lou tribe knock out their four central lower front teeth for reasons that no one can explain to me except as 'tradition'. Alfie had little leisure except on Sunday when he went to mass very early and entertained a very cuddly girl who might even have been his niece, for a short time in the afternoon. We grew very fond of Alfie and I know he liked us as well as the imbalance in our relationship would allow. It was a condition he made in feeing to us that when we left Kenya we would get him another musungu (white man) to work for. We whites paid about twice what the Asians paid for help and about four times what an African would have paid him. The project to educate five children was quite dependent on us. That was really brought home to me when I lost the St Christopher that our children had given me when we left for Africa. Fiona had hers torn off her neck at a traffic light within a month of her arrival in Nairobi but I succeed in keeping mine for a couple of years. Somehow it had come off in the bed and I told Alfie to look out for it. He couldn't find it. This made me a bit cross and I said, 'Come on, Alfie, we're going to find it – now.' And we stripped the bed carefully but the St Christopher just wasn't there. It was worth about £100 and Alfie knew it was valuable. He thought he would never work again if it didn't turn up. Our bed in Nairobi was a wooden affair, planks like floorboards with a foam mattress laid on top. We had given up when, down between two of the planks, I caught a gleam from the medallion. When Alfie saw that it was found he fainted, just momentarily, but he fainted. He didn't quite fall down and he was all right after a minute or two, but it was a sobering moment for me about how ill-divided is the world's goodness.

My favourite memory of Alfie was his polishing the floors. He put on animal skin moccasins and skated about on the floors, shuffling his feet madly, and brought up as great a shine as any electric floor polisher, the little wiry body bent to the task and beads of sweat standing out on his head, which was no more than five feet two inches off the ground. The last time I saw the little man he was still smiling, but the odds had beaten him. He had spent nearly everything, including our parting gift of a cow, on educating his children, but there had never been enough. Typically, he would have had enough for two to go to school and they would have done their best to teach the others what they had learned. But now they were all in their twenties and thirties; one who had had a good job as a prison warder was dead and none of the other four had a job. He would never have known hunger, but Kenya still offered Alfie a hard life.

The great Char Les (I never got his surname) was another poor African who taught this peeling nose a bit about life. He explained the tradition in East Africa that if a landowner is not using a piece of land anyone can come along with their jembe and dig herself (or himself) a vegetable patch. Similarly, if you have goats and you come upon unfenced land, you are entitled to let the goats help themselves. The only snag is that if the owner wants his land back you have to remove your crop or your goats as soon as practical. All over Nairobi we saw pieces of land being cultivated. My favourite was the lady who farmed a long strip between the dual carriageways on the main road north to Thika and the pineapple fields. These informal 'shambas' looked very insecure to me. How does the squatter make sure that it is she and not another of Nairobi's teeming millions who harvests the green beans she has sown? 'No. They cannot take,' Char Les tried to assure me.

Char Les was the gardener who looked after the compound in

which we had our first house in Kenya, a flat overlooking the Kerichwa Kubwa River, a grey burnie which gurgled its way down to the join the Nairobi River. Just across the unmade road from our compound were some open woods and Char Les and I decided to have a shamba there. I would provide the seeds and Char Les would provide the tools he used to cultivate the grounds of the compound. I would teach Char Les a thing or two about work, I thought though, as it turned out, the settler was the one who learned most. Like a settler of old, marking out his slice of the untamed bush, I stuck pegs at the corners of a bit about fifteen yards square. I started in one corner with ambitions to get Char Les to work alongside me while we gradually worked our way across our shamba. But Char Les wandered into the middle and started digging. When I tried to get him back in line, 'Come on, Char Les. Get in line and stop scratching about like a hen.' He refused on the grounds that the settler didn't know what he was doing with an African spade and he would be safer farther away . . . and as we were going to do it all what did it matter where he started? I let that pass and kept to my line. But as the work progressed and as the dark, sappy loam of the virgin land was broken in, I was to learn another lesson. Right on the border of the shamba there was a big boulder. It might have weighed two hundredweights but it was awkward, and I had to get it shifted off the shamba, as William Yull had done at Mains of Fedderate in the 1820s. I must have my field cleared.

Char Les was very interested. What is this white man with the red face and peeling nose trying to do now? Getting a little short of temper I explained, in patronising tones, that our field was to be square and we would need to get this big stone removed from the march. Char Les was astonished. 'But why, Bwana? If you shift this stone over to there you will be able to dig here. But then you will not be able to dig over there. In Africa we dig where we can dig. You can leave the stone here and dig over there.'

'For heaven's sake, man. Then we won't have a straight boundary.' The penny dropped slowly on the man who had lectured in Economics to the youth of Glasgow. The reason we

226

needed straight boundaries in Scotland was to accommodate our ploughs and our harrows, never mind our combine harvesters. We also needed straight fences to ease relationships with our neighbours. But in this equatorial forest we had no ploughs, no combine harvesters and no neighbours. But we didn't leave the stone where it was. The white settler must have a straight boundary however daft, and his African partner was very patient.

It took us a couple of short days to dig our shamba and bash the clumps into a weedless free-running tilth and then we were ready for the sowing. I expected a battle over what system of hand-sowing to employ. I expected Char Les to go for the scratching hen principle, but to my surprise he was quite happy to go along with the peeling nose's idea of rows; his only contribution to the plan was that we must plant the seeds about ten inches apart, whereas I thought five would be about right. Having won the battle of the big stone I let Char Les win that one. In three days, with the heat and the moisture in the rain forest, we had a healthy breer.

I went several times a day into the forest to inspect my crop. It grew so quickly I noticed a difference every morning. Then I had to go away for a few days. When I got back things became clearer.

I went at once across the unmade roadie and into the forest to inspect the crop. It looked fine. But what was this? The rows were different – thicker. Every second plant was what I thought was a strong seedling tomato. Although the compound was always well kept, the flowerbeds weeded and watered and all rubbish swept up, I had often seen Char Les, arms aloft and fingers through the netting wire of the high fence, a cigarette in his mouth as he gazed across the Kerichwa Kubwa Valley to the enormous acacia trees beyond. He seemed deep in thought, hung there for hours. I thought him a philosopher. But gradually I put two and two together. The little tomato plants were in fact cannabis. As far as Char Les was concerned my beans were just a nurse crop for his pot. Pot being illegal, not that anyone cared much, and being a foreigner in a strange land, I took one crop of delicious organic green beans and left Char Les to it. I often think of our partnership when I hear of some peeling-nose with a plan for saving Africa, and try to guess what will be the African agenda.

We really had a great time in Kenya. I will always feel warmly about Chris Mackel, the grain expert who used to tell my listeners to *Scottish Farming News* about the grain market, for his kindness in sending me a letter in which he told me all the latest news from Scotland – all the news that I would be missing in the jungle. But Mackel totally misunderstood what it would be like in Nairobi. There was no need for anyone to feel sorry for us. We had a nice house near the centre of town and a garden that grew more fruit than we could eat. As well as the avocadoes, we had enough bananas to feed the whole of Methlick, we had mangoes, cape gooseberries, guavas and several fruits we had never heard of including delicious super-sweet loquats. We discovered muglai curries produced in fifty excellent restaurants by the descendants of the Indian labourers brought over by the Raj to build the railway from Mombasa to Lake Victoria seventy years before. And swilled it down with excellent local lager at 50 pence a pint. The equatorial sun shone down on us, but, as we were at over 5,000 feet, it was never too hot. And never mind the international rugby scores which Chris so kindly sent us; the first *Daily Nation* newspaper I read contained a report of the Second Division match between Stenhousemuir and Forfar.

The loon from Aberdeenshire played cricket to the highest standard and on the best grounds he'd ever played on. One of my first games was at Limouru Club at 8,000 feet in the Kinangop hills among the coffee and tea plantations. I opened the batting for the St Andrews Society against the local White Settlers' club. It was a sort of Burns cricket match. We were piped out to the wicket. Then we had to drink a dram with the piper. As the piper was preparing to leave, the home captain said, 'Hold on one minute, piper, and you can pipe one of them out again.' Among those White Settlers I was not that far from home.

With Alfie keeping us clean and very well fed, and Joffrey keeping the garden and the car, the sun shining every day and thirty restaurants still to explore, we had everything we could want. So why on earth did we come home?

Well, it was my fault again. Fiona would have stayed. She got on really well with her black colleagues, with whom she had a great tea party every day at ten in the morning. They told her about being brought up in the villages, chasing down and cooking antelopes, and taking their trousers off when enjoying mud slides in case they would get a thrashing if they came home with muddy breeks. On one occasion she jumped up to make some more tea when the pot went empty and this had led to a bit of eyebrow-raising whereupon, in a remark she cherishes still, her boss, a very clever young Kikuyu, said, 'Allan is not a typical Scottish.' They wanted to keep her; they would even have got me a job at Nairobi University just to keep Fiona.

But I wanted home. I didn't want to become just another grumpy expat complaining in a strange land about the standards of catering on Kenya Railways. I was told by an expat when we first went to Kenya that you only needed three words of Swahili; jambo (hello), asanti sana (thank you) and bugger off. After three years I still thought that was pretty depressing advice to give a newcomer, but I was beginning to see what he meant. And I didn't want my time left in the driving seat of the family farm to be spent 7,000 miles away. Besides, if I were to be able to pass Little Ardo on to another of the family that has farmed it since 1837, without lumbering him with Fiona and myself as overheads, there was great need, having got rid of the overdraft, to build up a positive balance to ease our old age. The Retirement Fundie had to be started.

CHAPTER TWENTY-ONE

Jean and John Allan RIP

1986 and 1991

Three Acres and a Cow
The Love Song of Hodge to the Princess
by John R. Allan

Princess, the moon shines on the dew
About your forest lodge
Where I, in amorous purview,
Stand mute, adoring, Hodge.

Within, the courtly revel goes
In stately minuet
And suitors stand in frozen rows
Of royal etiquette.

They offer thrones with studied art
Or empires with a bow
But I can only give my heart,
Three acres and a cow.

Three acres! They may seem a toy
Yet they've a fertile loam.
Just one young cow! But o my joy
If you would call her home.

These are not all. I have a mare,
Ten chickens and a dog,
Besides Belinda, round and rare,
Incomparable hog.

I have a house all gay without
And apple sweet within,
A small rose garden, walled about
That overhangs a linn.

These are not all. I have a song
Of love is near my heart:
A unison – and o I pray
For you to take a part.

So Princess leave your courtly rout
And fly away with me
To see the sands of time run out
Beside the sands of Dee.

Princess, who have the royal power
Of life or death to give,
Bid me to love and from this hour
Your husbandman to live.

Bid me to call you sweet housewife
And you shall have, I vow,
All that I have, my love, my life
Three acres and a cow.

I have no doubt my mother was the Princess brought up in the
relatively rich home at North Ythsie. My father's had been a very
different situation, the lovechild brought up by his grandparents
and left alone when he was twenty, with an education and £30.
His granddaughter Susie has pointed out that this man who loved
words for themselves and who was an expert at crosswords has

called his hero Hodge, which is almost John backwards. Yes, my father was 'adoring Hodge' and the Princess was my mother.

So the economic climate was favourable in the early '80s, and crofting was a better strategy, but the retirement fundie was soon to get a boost from a sad quarter. Jean and John Allan had left the North-east in 1977 to end their days in England's Green and Pleasant Land. It worked well for a time but all too soon they found that their Jerusalem had not been builded there. They became unable to cope. They had a housekeeper who kept the house dirty, kept the change and took their credit card home with her. They were too far from the family who could have helped. They both began to long for Scotland, and the old man to look out on mixed arable fields again. Most of all, they longed to be back in Blairlogie, where they had been so happy before the war and where their only son had been born.

Their retirement idyll had only lasted five years. In 1983 they abandoned their house and booked into the best hotel in Lincoln at a cost of £650 a week. That would have been expensive for a holiday but it was unsustainable as their new way of life. As soon as we heard about it Maitland Mackie, my cousin, and I took a Range Rover each down to Lincoln, made a bed for one in the back of each vehicle and took them to the Blairlogie Hotel. With tender loving care there they recovered marvellously and were soon into a very nice ground floor flat and garden just three doors down from Croft House, which they had bought in the 1930s and where I had painted my bike, the doorstep and myself a nice shade of pink in 1943.

They seemed happy there, with a very good neighbour who looked after them; and then suddenly, one morning, Jean found John dead in his bed. He had been attending the hospital for blood treatment, including a transfusion which had seemed a great success. No one used the word cancer, but that is what it was. I was on the point of going off to Kenya to join Fiona and asked his doctor if Dad would be all right. He assured me there was no need for me to hang about. John Allan could live for another three years.

He was dead within a week. My parents had been married for fifty-five years and if they ever had a hard word between them, I never heard it.

I don't know if I should have, but I went off to Africa after only a short delay and left my children responsible for their granny. John and his girlfriend lived in Edinburgh, where he was a struggling musician, and they took Jean's great-granddaughter to Blairlogie on several visits. But the main burden fell on the eldest grand-daughter and head of the family in our absence, Sarah Purdie, now the New Deer chemist. Sarah soon had a bungalow bought and Jean moved home to Methlick. With Sarah's help, and Florence Cadger (who had helped her in the house at Little Ardo) and Betty Gray, a kind neighbour, cleaning and giving her a cooked meal every day, even bringing the odd plate of her own home-made soup, Jean survived away rather well. At least I thought so when I stayed with her for a month when I was home fulfilling a long-standing engagement to sing for other people's suppers at Pittodrie House Hotel. We noticed that her memory was letting her down and I got our excellent family doctor, Ronnie Chisholm, to see her. He asked her who the prime minister was, and this woman who had been intensely interested in politics all her days, and who would have had a lot to say about Mrs Thatcher had she still been all there, found that very hard. You could see she was trying and then she burst out vehemently, 'That wifie!'

Ronnie diagnosed Alzheimer's and said that it would take five years. When I came home very late one night from my singing, a little voice from the bathroom summoned me. 'Charles, I can't get out.' The top had come off the bolt in the door so there was nothing with which to push it along. The poor old lady had been stuck in the lavatory at least since bedtime. She hadn't panicked – she was just waiting. But there was a similar problem to come, similar but much worse. It was just before we came home from Kenya. In 1989 Jean had gone outside one night, no doubt to visit Phoebe Lind who lived next door, but had somehow locked herself out in the middle of winter. The milkman found her in the morning. He roused the neighbours, who managed to get her

233

inside whereupon, and to their astonishment, the cold body gave a groan. Jean survived and by the time I got home from Africa three days later she was fully recovered and being a nuisance to the nurses at Aberdeen Royal Infirmary, where she refused to stay in bed. She was too busy exploring.

Jean never went back to her bungalow.

We got my mother into a very nice nursing home at Skene where, to my astonishment, she was very happy. She had led an intellectual life and would have hated the idea of ending her days in the circle watching the lowest common denominator of television. She could no longer read but she found something there she had not had since John died. She got a friend. The two old ladies delighted in each other's company. They held hands and they gossiped and giggled for happy hours together. And as Jean became less coherent in conversation the two took turns of speaking a bit of nonsense and then they both laughed, whereupon it was the other's turn. Some of the happiest times I remember with my mother were in the home joining in that intimate and jolly but meaningless type of conversation.

I didn't get on so well when I accepted Sister Cattenach's invitation to sing to the old folks. They were all assembled in a circle and most were awake. I had my guitar and was sat beside my mother, whom I expected to be proud and perhaps to join in, although she had been tone deaf all her life. But when I started to sing my mother was alarmed, or perhaps embarrassed. 'Shut up, Charles,' she said. I continued as best I could but she was distracting. 'Shut up, Charles. Charles, will you shut up.' And then 'Oh for God's sake! Charles, will you be quiet!'

Over a couple of years she just dwined away until one day in 1991 she was dead.

With the passing of my parents their estate was all returned whence it had come and, extraordinarily, in almost equal measure. I felt as many do, that the old people having escaped from the farm with a little money should be happy to spend the lot. But Jean and John Allan had in fact lived off their state pensions, the interest of their nest egg, and the very small and occasional royalties from

John R's books. When the bungalow was sold and their shares transferred, my retirement fundie had recovered all that I had given them for the title deeds of Little Ardo.

To my surprise both my parents had left instructions that they be cremated. But they are both remembered on the gravestone on one of the three plots so thoughtfully bought for his relatives by John Yull, my great-grandfather and the farmer of Little Ardo till 1905. When we were getting their stone engraved I thought it a good idea, as a job lot might be cheaper, to get my name put on it at the same time. Fiona did not like that at all so our options are still open.

John Allan and Jean Mackie had been a rare couple who may not have added much to the steading, but they did manage, by harnessing their great sense of humour to the job, to get the council to tar the Loans – the half-mile of council road that joins Little Ardo to the main road to New Deer. Jean melted those most glacial of hearts, those of the Roads Department, with a letter telling them of her fears that when she and her husband died the hearse wouldn't be able to get up the Loans to take them to their plot in the old kirkyard. The excellent surface they put on survives forty years on with only small patches, and all those trees they planted made a big difference to life at Little Ardo. Jean and John it was who transformed Little Ardo from a beautiful Georgian house on a bare hillside, with good land and a fine view, to the wooded thing that it is today. The first thing John Allan did when he came home from the war was to plant 3,000 trees to provide shelter for man and birds. He saw that while twenty years is a long time for a young man to wait to see the fruits of his labour in planting trees, the same twenty years is no time at all for an old man to look back and see the difference he has made. Well, John Allan had forty years to look back, and I know he was pleased with how he had clad the bare hill. One of my fondest memories is of seeing him gaze in wonder at the trees he had planted in the mid '50s.

But Jean and John's most important contribution to the story of the farm had been that they never sold Little Ardo. Maitland Mackie's other children sold their farms (although Maitland III did buy Westertown back). As farmers rather than landlords it made more sense to have money to put into the business rather than having it tied up in land. When my parents were wrestling with how to escape with some money, John Mackie, by far the most successful farmer among my mother's siblings, advised them to sell to an insurance company but to secure the tenancy for me. I don't know why they did not take that advice. With the escalation there has been in land values there is no way we could have remained here if my parents had sold. It is because they didn't sell that our family still has this place to call their place.

But John Allan's contribution to the north-east corner of Scotland is much more important and endures still. Since his death his iconic book *Farmer's Boy* has come out in two new editions, the latest being the tenth. An eleventh edition, an audio version read by myself, is under preparation and should be out in 2010. And his *The North-east Lowlands of Scotland*, a book so important to the Lowland Scot, was also reprinted in 2009 in a splendid new edition by a splendid old publisher, Yeadons of Elgin, who started a new episode in publishing with a reissue of John's masterpiece.

The deaths of my parents, in his eightieth year and in her eighty-fourth, were strangely unemotional events as far as I was concerned. They had done very well, but they had died in time. And yet I did shed tears, not really over them but over their funerals.

John Allan had been a supporter of the Methlick Church all his days in the parish. I well remember his disappointment at the poor trade the field he had given to the Kirk to rent out had made at the fund-raising auction. The grass on the Waterside Park had been knocked down to a very careful farmer. 'I didn't give that park so that Charlie Addison could have cheap grazing,' he moaned. But when I went to see the minister about John's funeral he said, 'John Allan never came to the church when he was alive so why should the church let him in in death?' John Allan was an agnostic with a great sense of humour and would probably have thought that funny. But his son wept tears of impotent rage. The minister of Tarves was a less principled man, though perhaps more Christian, and he let us take the old man's remains the six miles to the next parish. There the Revie, the old Tarves minister, now retired, with whom he had written his name in the snow when he was courting Jean Mackie at North Ythsie half a century before, conducted his service without dwelling too much upon the mysteries of his god.

On the other hand, when I went to seek a venue for the last respects to my mother, there was a new minister in Methlick. There were tears again but this time they were tears of gratitude. The Reverend Angus Haddo told me 'As long as I am in Methlick the church will belong to all the people of the parish.' Jean Mackie was a poet, and one day her little volume of poems of old age will be republished, for they are strong stuff. This one could have been her epitaph:

> Once I was a young girl weeping on a rented bed.
> Now I am an old woman
> With two public rooms
> And three bedrooms, all empty,
> And I cannot weep.

Even now she is not 'little regarded' as she wrote of herself in another peom. In 2006 she was included in the *Biographical Dictionary of Scottish Women* for her contribution to progressive education. She was some dame!

They were quite a couple.

CHAPTER TWENTY-TWO

Charlie Allan – Crofterman
1989–97

On this land I've made me own I just struggled on alone.
But it's nearly over now, and now I'm easy.

Looking back over the accounts of what I may call my first period
as the farmer of Little Ardo, from the time we came home from
Glasgow until I started at the BBC, it is clear that the lack of
balance between ambition and funds is what had most bedevilled
the enterprise. I might have been able to put up those sheds, grain
towers, handling pens and Dutch barns, and laid another acre of
concrete – I might even have bought more acres – if the exotic
cattle boom had just lasted a few years longer, or even if it had
declined gradually instead of in an all-in-one oil crash, as it did in
1973. But it is clear that two things doomed my ambition. First was
the cost of borrowed money. In the year of 1978 for example, I had
a trading profit of £17,000 but the bank needed £35,000 in
interest on the overdraft and there was another £10,000 to the
mart for the cattle they financed. And the other sticking point was
the cost of labour. It wasn't that I paid too much, just that I earned
too little for a payroll that peaked at three plus the farmer. This
time there would be no staff and at least I was starting with no
overdraft. I would borrow on a seedtime-to-harvest basis only and
so there would be no monthly disaster from the bank.

Outside the crofting counties, where a croft is a small piece of land completely surrounded by legislation, a croft is a farm where the farmer does all the work himself and has a job off the farm to make ends meet. I like the story of the politician who tried to tell a Crofters' Union meeting that what they needed to do was to diversify. That was met with incredulity, 'Is that not what we've been doing for years – marrying nurses and teachers?' Of course, I had always diversified in that way and Fiona would still go off to the toon, first as computer manager at the Aberdeen College and then as a contractor to help the Hydro prepare their computers from the forecast disasters for computer land of the year 2000. I had no plans to get another income off the farm but I intended to do all the farm work myself with contractors for the expensive machinery work and sons and sons-in-law for any remaining spade and graip work.

The signs that things would be better this time came at the end of harvest, when the grain merchants had paid up and the cattle were all away off the grass. Before we went to Africa I used to get what I called my 'overdraft window'. That was the short period after harvest when so much money had come in that my overdraft had a brief spell below its limit. But when the harvest of 1990 was banked I had a new and glorious 'interest window'. For a few short weeks the overdraft was paid off. Farming was going to be better fun this time around, I could tell.

And the crofterman of Little Ardo did get an income off the farm as well. Both the *Press and Journal* and the (Glasgow) *Herald* fee-ed him to do an article a week. When *Leopard*, the local heritage magazine, went out of business, I bought it and persuaded my daughter Susie, who was a reporter in that excellent nursery of newspaper talent, the *Dundee Courier*, to take the job of editor. Fiona wrote an accounts system, I sold the advertising and also wrote an article in it. What all that brought in wouldn't have restocked and equipped the farm but it certainly helped with the housekeeping.

But this crofting did imply a big change in status for Little Ardo. When John Allan came home from the war and bought it from Maitland Mackie he could have paid for it with his first 16 months' profit, produced by his six men and forty cows, while he watched with his foot on the gate or got on with his writing.

When I had been home from Africa a year, I was to write in my diary, published every Monday in the *Herald* for twenty years, 'I see myself as like a street entertainer who plays the mouth organ, accompanies himself on the guitar, has a set of cymbals between his knees, a drum he can kick and a set of castanets under his oxters. Certainly that is what always comes to mind when I am drying wheat, a lorry comes with feed and I can hear the phone ringing. I don't answer that because I know it is a neighbour to say I have cattle out on the road. Then I wish I had another pair of hands like the bandsman who wishes he could play the clarinet as well as everything else.'

So I was busy when I came back from my holiday in Africa. And it wasn't just a question of 'don't take on deadweight debt and don't take on staff'. I still wanted to have a beef herd and for that I went back again to the drawing board. The plan was another piece of fiendish cunning using transplant technology. I defined the perfect cow as one with the mothering abilities of a dairy cow but with easier calving, which would produce calves with beefy conformation and the ability to grow very quickly. My ideal cow would eat very little herself, cost very little, look pretty and have two calves a year. You will see that the Jersey cow is nearly there. The biggest snag is that Jersey calves, even if sired by Charolais bulls, are slow-growing beasts with a very low proportion of their meat being roasts and rumps. Sausages and mince were not what the market required. But what if we transplanted beefy embryos into Jersey cows? The calves would get all their mothers' care and attention without taking on their mothers' shape. Now, just at this time Dr Peter Broadbent of the Scottish Agricultural College was developing a plan to take embryos from the best fat heifers at Buchan Meat's slaughterhouse, fertilise them with semen from beef bulls and transplant those into dairy cows. Better than that, they were to transplant two embryos into each cow, which would keep down the size of the calves to help with calving. Those cows would indeed produce two calves a year.

It was a brilliant idea, even if it didn't work. Soon I had a herd of Jersey cows which cost about a third of what I would have had to pay for anything better. They looked lovely on Ardo's braes. Broadbent soon had them implanted with two Limousin embryos each and then we ran them with Argus, the Simmental bull, just in case any of the transplants didn't hold. That was not in the mistaken belief that a cross-Jersey calf would be better than nothing, but because, in order to qualify for the beef cow subsidies, you had to have decent calving percentages. The transplanted calves were good beasts and the Jersey milk made them grow like mushrooms. But there weren't nearly enough of them. There was not one single set of twins. And the half that didn't hold on to their transplanted embryos embraced Argus with enthusiasm. We ended up with just under half Limousin beef calves and the rest slatey-ersed cross-Jerseys. And the little Channel Islanders weren't fit for life under my husbandry. In 1991 we had 15 per cent deaths from milk fever and they couldn't compete at the trough with the black Hereford and Simmental cows. I had dreamt up another brilliant failure. I bought a load of black Hereford heifer calves and set about creating a more robust hill-cow herd.

The change in Little Ardo's circumstances wasn't the most important development in agriculture, for this was the era of the MacSharry reforms, it was the age of set-aside. At first we expected disaster but, because they got it all wrong, the MacSharry reforms meant that, for a year or two, just like the period after the war, any fool could make money farming and at least one did. It was the era of set-aside, but it was also the era of my retirement fundie. I added to my wish list for the farm, that Fiona and I would manage to put away a sufficient fund that we could leave the place to the next tenant without the huge burden of debt which had made our time there so hard to enjoy.

Ray MacSharry, the European Commissioner for Agriculture, aimed to reduce the cost of running the Common Agricultural Policy. That policy had been to guarantee farmers a good price for as much as they could produce. The Commission bought any surplus on the open market at attractive prices. The trouble was they were too attractive. Soon we had huge grain mountains and wine lakes in intervention stores. Because the prices were so high farmers got better and better at producing and every acre of marginal land was ploughed up to produce food no one could afford. As Neil Godsman, the farmer of Cairngall, put it, 'Aabody fae Rothiemurchas tae Outer Mongolia plooed up anither parkie for grain.' It was said to be bankrupting the European Union. MacSharry would tackle the problem directly by forcing farmers to set 15 per cent of their land aside. That is to say, to grow nothing on 15 per cent of the land on which they had been growing cereals.

The farmer of Little Ardo was scandalised. I wrote in my diary, 'I am determined to grow something on my set-aside. My ancestors demand it. They did not drain and de-stone this land, nor did they carry seaweed from the coast to fertilise this farm so that I could grow weeds. God knows I grow enough of them already without reserving a seventh part of my inheritance exclusively for weeds.' Some exceptions were allowed. Hay for horses would be OK, and a laird in the south had got permission to grow grain provided it was fed to pheasants. I considered doing the same but feeding it to rats. That would make more sense, because rats being inedible wouldn't be adding to the food mountains.

But we were wrong to worry. Set-aside meant being paid for going back to the old system where one field in six was left fallow each year to let its fertility recover. Set-aside, as Ian Davidson the Guru of Grain, who was to play an increasing role at Little Ardo, put it, 'is just £150 an acre for growing a perfect break crop between cereals'. Now, what the bold Ray expected, and so did everyone else, was that farm prices would fall when the support of intervention was removed. The acreage payments, which varied from crop to crop but were about £100 an acre, were to be compensation for those lower prices. But what actually happened was two things, both good for the retirement fundie. The setting aside of all these acres led to shortages so that the prices went up instead of down. So we were compensated for a fall in prices which didn't happen. And then there came Black Wednesday, which we farmers call 'Blessed Wednesday'. 'Peerie Norrie' Lamont, the

Chancellor of the Exchequer, came sheepishly out of No. 11 Downing Street to face the world's press and announce that Britain was to leave the European Exchange Rate Mechanism. That meant a devaluation of the pound which raised British farm prices by another 15 per cent and also raised the value of all those compensatory acreage payments.

So the thrifty crofter was making profit at last. Thanks to Ray MacSharry and Peerie Norrie, he could hardly help it.

I had another try at keeping pigs, this time under Mossie's guidance. We modified the piggery for fattening loose-housed pigs. That cost £300, the same as Old Maitland Mackie had spent in 1935, building and fitting out with small wooden pens all 146 by 36 feet of the original shed. The first batch was 200 weaners on contract for Grampian Pigs. That brought in £850 for ten weeks' effort and seventy-five big bales of straw. Two years later I was to sell seventy-five bales for £1,050, which was a lot easier than processing it through pigs, but even so, it was still a profit. I was able to install what I had failed to get with my barley beefers – a fully automatic system. At under £500 I got a second-hand bin and all the augers so that the feed company sent a lorry which filled up the bin, and the augers delivered it to the feed hoppers without the stirring of a human hand. As soon as a pig took a mouthful there was a gentle whirring noise and feed was delivered from the bin to refill the hopper. I used to like listening for the noise. When it came on I could say at last, 'There's Charlie feeding his pigs.' How wonderful, but it was not foolproof.

My first two batches of porkers were on contract to Grampian Pigs. But the third lot I financed myself. That was consistent with my plan to avoid deadweight debt. Seedtime to harvest with these weaners was only ten weeks. Two things happened to spoil the equation this time. One was that whereas with two lots of visiting pigs, four hundred in number, not one died; when I went into business on my own I lost eleven pigs – 5.5 per cent. Worse than that, the visitors had converted pig meal into pork at the rate of 2.8 pounds of feed to one pound of pork, but my own pigs converted at three to one – not nearly so good, but why? Was it disease? Were the feed company swicking me with inferior feed? Or was it those deaths? I noticed another strange fact during that ten-week cycle, and it was not unconnected. We had an Ayrshire-cross Shorthorn cow who did unusually well. She took on a sleek appearance and her coat shone like a show beast. Now, the cows had been going in and out all winter and the feed bin is on the way to the braes where they picked about for daisies among the snow. One day, after the feed lorry had been, I saw my cow licking the concrete under the bin. 'Ah, the delivery man has spilt some nuts,' I smiled to myself. But my clever cow soon wiped that smile off my face. When she had licked the platter clean she gave the pipe a smart butt with her nose and a shower of nuts cascaded down on her from the joint of the pipe and bin. I soon had that bin fenced off and the pigs' conversion rate recovered.

CHAPTER TWENTY-THREE

Mossie Points the Way Forward

1992–97

My first spell of farming Little Ardo between 1973 and 1982 had been interesting, and I had been proud to be filling the shoes of my ancestors, but really it hadn't been much fun as I watched the overdraft spiralling while my carefully laid schemes usually went agley and the others led to disasters. Oh, I did make a profit. My average profit was £800 a year. That was the sixth part of what I had been getting for a cushy life at StrathclydeUniversity. I would have been much better off if I had been one of my own employees and the damnable thing was that I had to work fifty hours for my £16 a week. But now, after my four years at the BBC and three in Africa, now that I had become a crofter, farming from 1989 to 1997 with my own and my wife's money and my own sweat, and with MacSharry working wonders, I was able to average £13,000 a year. That was fun – and it got funnier and funnier.

The plan I brought back with me from Africa had been to get contractors to do all the expensive machinery work, a crops consultant to tell me what sprays and manure to put on and in what quantities; the buyer would select which of my cattle were ready for the butcher, and I would do, with occasional help from cas. lab., all the day-to-day work myself. Because the system was right, or not as wrong as my earlier attempts, the little farm on the hill made money each year between 1989 and 1992, though 1992 was a bad year for output. The cause was too much rain that year. There was some profit, but the beasts didn't do well. The grass was thin, the silage was wet and the crops were mostly very poor, though MacSharry's acreage payments ensured that they left a bit of profit.

Worst of all was the Tapidor winter oilseed rape. I had over-heard Mossie, the county's leading grain baron, saying 'Tapidor is the way forward for Aberdeenshire,' so I had jumped in. I was rewarded, not with the usual tonne and a half to sell, but a mere half tonne – with the added humiliation that the crop was on the brae face. Every driver between Methlick and Fyvie saw that field getting blacker and blacker all summer. The only thing that really throve in 1992 was the weeds. Towering above the wheat we had the finest stand of day nettles I have ever seen. That was the end of consultants. If it comes to growing weeds and letting disease run riot I can do it myself. Having got a truly horrendous trade for some of my slender Jersey-shaped cattle, I decided I would pick my own cattle for market. Employing experts is all very well, but there was no way I could see to avoid learning a bit about the job myself. Perhaps James Low was right when he told me that, instead of going to universities to work, I should have stayed at home and learned farming.

Aiming high as usual, I went for the top man from whom to learn. Ian Davidson, the farmer of Moss-side of Tarves, was the undisputed leader in growing combinable crops and he was in our discussion group that met each Sunday for a pint, to criticise the government and our neighbours, and to plan the week's work. Despite that disastrous tip about the Tapidor, my faith persisted. I knew Mossie knew a lot and my first plan was to listen. Mind you, that was not a simple job. Well, there was the Tapidor, but Mossie also liked to put snoopers like me off the scent. He knew everyone was watching to see what he was spraying on his fields and delighted in taking his sprayer out in winter and running up and down with just water in the tank. When he was asked how he managed to get his rape looking so fresh in March he told an admirer that it was 'green paint' he was putting on. And he confused a study group up from Yorkshire who asked him if his recipe for Cycocel, the straw shortener, was in pints to the acre or litres to the hectare. His answer wasn't worth the trip from Yorkshire but it was well worth coming the four miles from Little Ardo to hear. In case his guests were just getting a bit too nosey he said, 'I couldna say, really. We work tins [of chemical] to the fill [of the sprayer]'.

One spring when the frosts weren't even offering to break, his answer to the question of whether it was time to be spraying and what to put on was, 'Aye, it's time to get on the bluecol' (anti-freeze). So you could have made an awful fool of yourself following Mossie's advice in those days. There was a lot of sense there, but you had to be careful.

When I told the discussion group my idea that I would need to learn the job and do it myself, though I wasn't sure I could do it, Mossie must have been in an unusually good mood. He said, 'Ach, I'll keep you richt.' In the same way that he was tempted by the challenge of growing oats on a billiard ball, and had in fact grown a profitable crop of sunflowers in his moss in the wettest year we could remember, I think Mossie was tempted by the challenge of getting Charlie Allan to 'fairm richt'. I made the great man laugh a lot, and he really likes laughing. We both had very sore joints, and while mine turned out to be arthritis and needing surgery, his was a simple case of Tennant's elbow which was cured by drinking left-handed for a few days.

Mossie was really very good fun and part of that was his odd way of thinking in wobbly lines. Like so many Aberdeenshire farmers Mossie loves to gossip. Every shocking tale of goings on in Aberdeenshire goes through his hands and usually comes out even more interesting as a result. When I told him I was reluctant to tell him a secret of mine he was indignant when I told him I didn't trust him to keep my secret, 'It's nae me that lets things oot. It's the folk I tell that canna keep their mous shut.' And the pupil didn't need to be too thin-skinned. Mossie was not the gentlemanly coach that Professor James Fowlie had been. When James thought my stots weren't nearly fat enough for the market he moved the pipe to the other side of his mouth and said, 'Well, Charlie, if ye dinna get what you're expecting you can always take them home for another few weeks.' But when Mossie came upon my new system for loading pigs just after I had cemented in the uprights he shrieked, 'That's nae damned use', and charged in amongst the wet cement, tore the posts out and flung them to the ground.

The combination of MacSharry's miscalculations and Mossie's good advice made farming at Little Ardo a whole new ball-game. It was also a new game for the farmer, because it brought with it an honorary membership of Mossie's Gang. That was a new experience for me in many ways, including that I had never had friends who were younger than me. There had been exceptions of course; I wasn't an ageist, but it had started with being so precocious at sport. That meant I played football in men's teams when I was fourteen and cricket at fifteen, so my best friends were older. Then my girlfriends, throughout my life, were always a couple of years older. That's nothing between Fiona and me in our seventies, but at sixteen two years either way is a lot. Perhaps they felt safer with a mere boy. Sport also threw me in with the older students at the university. When, on university staff, I became a champion at the Highland Games, I became a fixture at meetings with visiting top (and therefore older) brass so that the host could change the subject to something entirely different by getting me to explain how tossing the caber was judged. It made a bit of light relief from endless gossip about promotion and university development. But Mossie's lads were almost all younger. They were different and they were amusing. They had largely grown up together but now they were the men who would still be boys but who were more dangerous because they now had money. The excesses which the gang got up to are pretty well documented in my *Farmer's Diary* books, but I can give a flavour here.

Mossie led a great team, almost the entire first eleven, to Smithfield Show in 1993. There was the usual jolly excess all day, culminating in dinner. That was always special, and extra special was Keith Nicol's birthday dinner in a Chinese restaurant in Soho. We had been there before, but despite that the genial oriental host was pleased to see us. So pleased in fact, or perhaps because he remembered the noise we had made the last time, he put us in his private dining-room upstairs. I was pleased that we were going to the China Plate because, having been there before, I knew that you could get all it was reasonable to eat, exotically presented and clean for a third of what we might have paid elsewhere.

Our host offered us his banquet at £7.50 a head, or his special banquet with a bottle of rice beer thrown in for £12. Oh no, but that wasn't enough. John Bain, the Tarves butcher, took charge. He told Mr Choi that the dinner was to be very good and it was to cost £18. Mr Choi was persuaded and went away to harvest another acre of bean sprouts and kill another hen. Well, it was quite obvious that we would have had more than enough with the £7.50 meal but poor Mr Choi didn't have a lot of extra specials he could sell us. The only way he could deal with our request for a bigger price was to bring us bigger portions. This feast was brought and we set to work while Mr Choi got on with the rest of his trade. Then Keith Nicol wanted to buy us all champagne as his birthday treat. He tried the waitress for discount on her finest Moët Grande Cliquot. She couldn't act without Mr Choi. But Mr Choi was never appearing, so suddenly Nicol jumped up saying, 'I'll get him', and disappeared downstairs. He reappeared in half a minute carrying the owner under his arm. I don't suppose he had ever been asked for discount in quite those circumstances. However, the discount was given and Mr Choi was so delighted by the whole life-expanding experience that he even baked Keith a birthday cake with 'Happy Birthday Keef Nikle' inscribed in pink icing.

The excess continued the next day with a shopping trip to Harrods (where else?). I don't think there was much bought but the lads just wanted a look at how the even richer half lived. We didn't see any Arabs with plastic bags full of £50 notes but we did see some astoundingly dear stuff and at least one purchase was made. My young friends had noticed that I was not a snappy dresser, nor was I very trendy, so they decided to give a fillip to my image with a Harrods haircut. At that stage in my life I had never paid £3 for a haircut and still cherished the days when Est-Alec at the Garage in Methlick gave me a crop for 6d. So it was clearly going to be a change. There were eighteen in the team and it was estimated that a whip-round of £2 each would do it. In my diary I wrote that my hairdresser at Harrods was, 'A very engaging young lady who wanted to speak about the political situation in South Africa and had no idea what a wet year we'd had, washed my hair first. Then she cut it, blew it dry, spiked it up in front, and sent me out to face the boys looking the way Bill Haley looked thirty years ago, but with the quiff standing up instead of curling down.' I gave a tip of £2, which was about what I usually paid for a short-back-and-sides in 1993.

The glitter of Harrods was just a warm-up, for Mossie had a splendid day out planned for us the next day. In 1993 there had been a bit of a financial crisis and in London all the whiz-kids were finding suddenly that there was more to earning a living than drinking champagne at lunchtime. One consequence was that the second-hand car market was swamped with expensive toys. There was to be a grand sale of Jaguars, Rollses, Ferraris and heaps of Mercedes at Blackbush, some twenty miles out of town. Mossie hired a bus to take us down. The top car was a custom-built Italian job that had once belonged to Rod Stewart, and when it sold for a mere £61,000, it was clear that bargains were to be had. The lads started scooping up those grand cars. The change in the financial fortunes of the farmer of Little Ardo were emphasised when, emboldened by the £38 haircut, he bought a less than two-year-old Jaguar for less than £9,000. That was more than three times my previous top price for an automobile purchase. And my young friends rubbed in how far I was out of my usual depth when it came to the tip for the driver of our minibus when we got back to central London. Somebody said 'a tenner', which I thought was reasonable, though I couldn't see how that was going to be divided between the fifteen of us. I only realised slowly that what was meant was a tenner each. That came to £150 for half a day's work which had already been paid for by her employer. My new car was not a success. Fiona all but refused to get in it. She said it just wasn't me. She was right. I kept it for a couple of years, got most of my money back and returned to my philosophy of getting a car which would take me as cheaply as possible from here to there.

Other little extravagances were made possible by the fact that the farm was paying at last. I took up cycling, and to make it as easy as possible to generate as much speed as possible I bought a very expensive fibreglass and titanium bike. It didn't cost as much as the Jaguar, but only my Isuzu Trooper had cost more and I reckoned I needed four-wheel drive and a tow bar if I were to do the job properly. Still, my bike cost me £2,750 and was my third most expensive vehicle ever, and yet it was money well spent. That bike enabled me to rescue my circulatory system, which had almost given up. On all but the warmest summer days my fingers drained of blood and turned white. On cold days I had to work hard to get any blood below my elbows. Now, you might think that would be something to do with a lack of blood pressure but the doctors said it was the opposite . . . I had high blood pressure. So I was encouraged to cycle as much as possible. Soon I was travelling all over Scotland every weekend and sometimes midweek to time trials and cycling like a lunatic. It was wonderful. I have always enjoyed the mindlessness of sports competition and now, at the age of fifty-four, I had found a new sport. The zenith of my cycling came in my sixtieth year, when I did a sponsored challenge in aid of a school in Masailand in Kenya. I would cycle a hundred miles in five hours without an engine or a break. I did it in four hours and thirty-nine minutes, an average speed of twenty-two miles an hour, and in the process came forty-first out of a hundred and fourteen in the Scottish championship. I raised some £5,000, which was enough to build a spanking new administrative block at the school in the Masai Mara. It had four rooms.

Other extravagances of the now well-to-do farmer of the little farm on the hill included golf. With Mossie, his big brother James and Ernie Lee I joined the new club set up in 1995 in the grounds of Meldrum House, the seventeenth-century tower house at Oldmeldrum. We formed a company called *braesogight.com*, whose sole function was to enable a corporate membership, which cost almost £3,000 a year, less than £750 each, and proceeded to train for the Walker Cup. Things were going quite well, and from a standing start we were soon all threatening to break ninety sometime soon. But it all came to a shuddering halt. The professional, Ian Marr, offered me a coaching session in which he would film my swing. Then we would look at Tiger Woods' swing. He would play the two of them back to me for comparison. Then he would superimpose Woods' swing on mine and I would see where I was going wrong. Well, there was just no comparison. When he tried to superimpose Tiger's swing on mine the two swings never even touched. I gave up. The swing is the thing. If you have a swing you can develop as a golfer, but without a swing you are wasting your time. You might as well try to coach the world's smallest man for the world's strongest man competition. If you haven't got the equipment it can't be done. My swing was like I was hacking logs for the fire. No. I was used to being good at sports; I was still not bad at cycling. I would stick to that. Ernie, who has farms all over the place – Methlick, Tarves, Poland – only managed to play five times the first year so his golf cost him £150 a round. The next year he only managed twice, so his golf was costing £385 a round. On the last year he didn't find time for a game. That only left Mossie, and he wasn't going to play without his pals, so the company folded and the membership lapsed.

There was lots of fun with Mossie's gang and not all of it was as expensive as that. I went all over the place with Mossie and his mobile barbecue business as a carver and court jester. We got a free day's fishing on the Spey for some reason and on that occasion something of the philosophy of my crops consultant sneaked out. The team was myself, Greig Henderson (then of Nethermill) and Mossie. Greig's son was a promising footballer and was playing in a big game that evening, which meant Greig would have to drive later, so he said he would drive us so that Mossie and I could have a drink. But I had a big day the next day, when I was due to compère a concert at Braemar, so I wanted to take it easy. I volunteered to drive so that Greig and Mossie could have a drink. But Mossie clinched the deal in favour of hedonism. 'Na na, boys. I'll drive and then we'll aa get a drink.'

But it wasn't all beer and skittles.

The reforms that Mossie brought to Little Ardo included

changing from manufactured nitrogen to urea, which was cheaper and didn't leach away if the weather changed. That meant I could get it on early so it was ready and waiting for the weather. He introduced me to a new contractor, Davie Jaffrey, who revolutionised the sowing of the crops by employing the technique known as 'plough and plant', which allowed us to sow on whatever day we chose instead of waiting, sometimes for weeks, for conditions to be just right. When you asked Davie when he could come the phone went quiet. The first time I thought he had been cut off. 'Davie? Are ye still there Davie?'

'Ay, I'm thinking.' After maybe another minute of complete silence he would say, 'I'll be on Thursday mornin.' His team would arrive on the appointed day with a four- and a five-furrow plough and his sowing machine would follow fifteen minutes later. The ploughmen never got more than fifteen minutes ahead of the sower, so the weather didn't matter. Be it never so wet above, the ploughs turned up a nice dry tilth. And be it never so dry there was always some moisture in the newly turned soil. I just don't understand why in 2010 so many people are still sending one plough out in November to get ready to sow in March. Perhaps it is that they can't think what else to do with their ploughman all winter.

But the most important part of Mossie's revolution at Little Ardo concerned the chemical policy among the combinable crops. The system had been that the adviser came round and looked at the crops. Then he wrote out a ticket with his recommendation. That would involve a cocktail of chemicals and he gave, beside each chemical, the name of the firm that had offered the consultant's clients the best price. I then phoned each of the firms and ordered what was recommended. A few days later those chemicals might all have arrived and I would phone the contractor. He would come when he was ready and the weather was right, and spray the fields.

There were always a few days between recommendation and action. A week might have been typical, and at least twice the crops adviser was back again with his next recommendation three weeks later, but before his previous one had been acted upon. That would never have done for Mossie. He inspected the crops and made his recommendation in 'tins to the fill' so that the sprayer operator, who had a poor reputation, couldn't make a mess of it. I then started the sprayer filling and went straight to Oldmeldrum to the Chemical Spraying Company's store and bought the chemicals. Then I came straight home, filled my sprayer and put it on. Sometimes Mossie would say to leave it until the afternoon or even till next morning, but timing was nearly everything. Many's the morning the phone rang and an angry voice would say, 'Have ye nae got that sprayin deen yet?' After the first few times the answer was always 'Yes'.

All this was a long way from my early days as a farmer, when I followed James Low's advice to make a good seedbed, sow and roll the crop and then shut the gate till harvest time. Indeed, my attitude changed from rather despising crops as much inferior to livestock, when my new crops adviser promoted the little farm on the hill to the forefront of crop husbandry. Mossie has made a well-merited name for himself as a pioneer and he used Little Ardo as one of his trial plots. One of his great steps forward has been the blending of crops to get bigger yields, the most obvious and one of the first being to take a six-rowed barley and blend it with the higher-yielding and heavier two-rowed varieties, because the crows couldn't get the hang of the six-rowed they left the blended crops alone. Following the great Croptec event in 2009 when all the world and his wife came to Moss-side to be shown the virtues of blending, everyone is doing it . . . even the college scientists are catching on. But we grew one of Mossie's blends of oilseed rape at Little Ardo in 1993 and other blends every year thereafter. The idea was that a short, sturdy rape kept its lanky partner up to the sun without competing with it for light. The shorter crop didn't do that well in the half-light below, but what it did yield was a bonus.

Now, it is true that 1993 was a much easier year than 1992 so the comparison isn't really fair, but the fact is that our combinable crops grossed £34,135 in the last year of the old regime and £65,520 in Mossie's first year as Tyrant-in-Charge. Better than that, and confounding what everyone had told me, I didn't need more

254

chemical to do it Mossie's way. I needed less. And I didn't need more manure. I needed less of that too. Instead of looking for three tonnes of barley I began to dream of four tonnes. I made it too, on individual parks, but I never did quite make an average of four tonnes. The four tonne average had to wait for the next farmer of Little Ardo. But my reign finished with five years of profits that varied from the 'nae too bad' to the 'pretty damned good'. Even the seasonal overdraft disappeared and the Retirement Fundie grew. A comfortable retirement beckoned, even after Fiona gave up her career as a systems analyst. Having dealt with the horrors of the year 2000 (the threatened 'millennium bug' never really materialised) she retired, thankfully, when the Hydro merged with Southern Electricity, who took over the computing for the new Scottish and Southern Electricity.

And how much did my consultant cost me? Well, apart from a year or two of my life, the only direct cost was that I had to pay for a day's shooting for the two of us. Now, I had shot all my life since I was about eight, when Old Jack Martin taught me to shoot with an airgun in Glendevon. Then I got my own airgun on the farm to thin out the sparrows in the dairy, where they were such a hazard to open milk cans. As a teenager I shot pigeons with a 410 shotgun, plucked them and roasted them, and my mother used them as the first line of defence of her larder when I came in late from the dancing. Then I had shot pigeons to keep them off the new-sown grain in spring-time, and crows that swarmed on lodged barley as harvest approached. I had been to the tenants' pheasant shoot at Balnaboth in Glenprosen and those hare hunts at the Spott. So I could shoot all right, but I had never been at an upmarket commercial shoot.

By the time we had paid for the brandy and port to settle the stomach in the morning, and the jolly lunch for the small army of beaters as well as shooters, and as much Stilton and port as there was time for, and the tips to the gamekeepers, and the fees to the laird, my good advice wasn't as cheap as it had sounded when Mossie said, 'Na na. I'll just take a day's shootin for the twa o's and you can pey.' I would have been happier being paid as a beater, but that wasn't the deal.

You may think that the mixture of twelve-bores and seventy proof was dangerous. I guess it was a bit, though I never saw even a near miss. There was the time one hero's wig was whipped off by a twig and somebody shot it in flight. I wasn't at the shoot in Kintyre where the bag was twenty-one pheasants, seven hares and one greenhouse. Nor was I in attendance when Ernie Lee, a very prominent leader of the gang, got a pellet just below his eye. Now, there is an etiquette in these matters which some of us think is important. If you shoot anyone you are expected to leave the field before the police get there. But on this occasion all that happened was, well I don't know if it was a protest, a joke or a bit of both. As usual at the end of the shoot the game was laid out on the ground; twenty-five brace of pheasants, three brace of partridges, four woodcock and ten brown hares. So in the gap between the birds and the hares Ernie lay down, still with a thin trickle of blood running from the small wound below his eye.

There was another piece of shooting etiquette which led to trouble, in the days when the mobile phone started to make life a so much worse for those around people who are slaves to it. The gang made it a rule that when shooting and when golfing, phones were to be shut off. But at one of Mossie's clay pigeon shoots someone, who knew perfectly well that he shouldn't do it, started bawling down his phone. 'No, no. You want to do business with me you get it there by tomorrow morning at eight o'clock . . .' Then Ernie snatched the phone out of his hand and threw it high in the air, whereupon Mossie atomised it with both barrels. What a blow for civilisation!

Now I must introduce you to the concept of the sook cycle in farming. There is an old saying that a good farmer spends money as though he is going to die tomorrow but cares for his farm as though he is going to live forever. However hard-up the farmer is, the farm should never suffer. Well, sooking is really just the opposite philosophy. Here a farmer is taking as much money out of the farm as he can without a care to how it will affect the fertility of the place in years to come. A lot of sooking went on when farms were let on short tenancies and worst of all was a 'course of cropping', which might be grass on which the tenant got to grow four years of crops and then hand the land back in grass again. Lime and dung were not used in a course of cropping. It was just a case of 'sook what you can and run'. But there is a higher order in these matters. There is the sook cycle. Thus, the old man stays on far beyond his time. He won't spend a penny on mending the roofs, renewing the fences, or anything. Then at last the young man gets his turn of the big bed in the front room, takes a wife and spends every hour God sends and every penny he had saved on seeing to all the things his father neglected. For that brief spell and at no other time the farmer may be said to be happy. But soon the overdraft starts to bear down on him. He economises. He stops buying young breeding stock but tries a few more years with the old ones. He won't renew the roans on the steading. No lime is bought and no compound fertilisers applied. His crops just get a jolt of neat nitrogen – enough to give another one crop.

Part of Mossie's philosophy is that if things aren't nice you'll just get fed up. And whenever he saw that there was a pane of glass needing renewed, or a bit of Coburn track that was squint he would say, 'Of course, you're sooking just noo.' That is the way. The farm will be sooked until another young man takes control in a blaze of optimism.

I hope you don't think that the author sooked Little Ardo before he gave up control to young Purdie, his successor. Perish the thought, though there were some elements of the sook. I had taken over at Little Ardo intent on modernising the place. I had done a bit, but there was plenty still to be done. I judged we still needed one large multi-purpose shed. My grandfather's dairy byre, my great-grandfather's barn and my great-great-grandfather's byre all needed new roofs. But those were slate affairs and the timbers would never have stood reslating. And while all but the original byre might have been some use with a new roof, the old byre, the same one James Low had said was done in 1930, was still done in 1997, but was also too narrow to take modern machinery whatever you did with the roof. Basically the farmer had at last the money to finish the job to his satisfaction, but he didn't dare to clear those historic buildings, fearing a thunderbolt from his ancestors. So he took refuge in an excuse. It was, 'The next farmer won't like whatever I do so I'll do nothing for a few years. The sook goes on.' It was that or drive a hard bargain and land the next generation with a mess of debt. Little Ardo was too small to be economic as a farm now, so it was going to have to be an attractive way of life – and farming isn't much use as a way of life with a banker on your back.

I hadn't left the universities and put on my boots until I was thirty-five. That was a year later than my grandfather, Maitland Mackie, took his off for the last time. From that age Maitland Mackie would do his farming by telephone. John R. Allan only tried his army boots once at Little Ardo, in the harvest of 1945, and was told by his grieve to stick to the office, where there was work that he *could* do. I wasn't keen to take mine off, even at fifty-five, though my knees were buckling with the jumping on and off tractors and particularly manhandling the heavy booms of my sprayer, I would have carried on, especially as there was a living to be made at it and the retirement fundie was growing. But I was desperate to fulfil the peasant's promise to his land – to hand it over to one of his progeny. The two boys weren't interested, and of the girls, Susie had married a folk musician who was dedicated to his singing carreer, which was to carry him away to America half the time; Sarah, the elder, had married Neil Purdie who, though trained as a ship's engineer, just loved farming and had been working for me for nothing. So the eldest lassie, with her chemist's shop for diversification and her enthusiastic husband, were the obvious choice. I kept the young man working away for nothing as long as possible, but about 1995 I noticed him starting to blaw about Little Ardo. A good sign but also a threat, perhaps. 'We're doing this', and, 'we've got that at Little Ardo', he would say. I could see it was time to be off. If I didn't grab this young couple then they would go and buy themselves another place and spend their youth on improving that and I'd have lost them. So the deal was done. They would build us a nice bungalow to our specification and we'd move out as soon as possible.

258

Having seen how lost my parents became as age overtook them in Lincoln, we decided to stick to the hill where the new farmer and his wife could defend us against the extremities of old age. They would be able to fight with tradesmen and there would always be a supply of grandchildren and great-grandchildren to bring in a pailful of coal in winter, and bring up the paper from the shop.

That is why, on 16 July 1997, I carried Fiona over another threshold. Having taken me into her rented dower-house at Banchory-Devenick thirty-seven years earlier, and after living in twelve other houses and flats, we were settling for good in a dower-house she had designed, our daughter and son-in-law had financed, and I had built for our retirement. We chose the highest point on the farm. It is stone-built, with local granite and Welsh slates from three of the fine farm buildings that were being replaced everywhere by bigger, roomier and bleaker buildings made of breeze blocks and asbestos. So the farm that had been acquired by the Yulls and taken on by the Mackies and then the Allans welcomed a new name, Purdie. The little farm had its fourth surname – but all in the one family. The new farmers are the sixth generation, the seventh is on the ground and, being a bit old-fashioned already, I'd say, easily ready to be producing the eighth generation of us on the little farm on the hill.

Postscript

So 1997 saw a new farmer into the old place. It was quite fitting that Neil Purdie should be a ship's engineer. He followed me, who had trained as an economist, and my father, who was a journalist. John Allan had taken over from his father-in-law Maitland Mackie, who was of course a career farmer, but Maitland had followed John Yull the auctioneer, who followed William Yull, the other dedicated farmer. It is just my theory, but I think Neil's passion to be a farmer had something to do with his schooling at Turriff Academy, a very rural secondary school, where a lot of the boys' talk in the playground was about their fathers' farms or the farms their fathers worked on. As a solicitor's son, the young Purdie had no hope in the bragging stakes in the playground against boys whose fathers drove big Ford tractors or Claas combine harvesters, let alone those whose fathers had taken prizes at Turriff Show. The fact that your father bought and sold all those farms wasn't nothing, but at Turriff Academy a lawyer's son was socially disadvantaged.

One of my earliest memories of Neil Purdie is of him being incredulous that somebody could go on holiday when he had a farm needing all that maintenance work done. Well, as a result of my sooking, he had plenty of maintenance to keep him nice and busy for a long time. He adopted a very sensible policy towards livestock. He kept on my hill cows, which were paying (though not much), and he took on boarding pigs for John Gyle, the go-ahead farmer of Mains of Esslemont near Ellon. He was paid so much per pig and Gyle provided the feed, which Neil collected from the Mains. Young Purdie took on just enough pigs to give himself something to do every day but also to yield a living wage for himself whatever teething troubles his farming might have. John Gyle liked Neil, which was quite a compliment, for Gyle was pretty fussy. In fact, seeing Purdie flying on by himself boarding pigs to give himself a subsistence, and driving his big cart the twenty-mile round trip to get feed and save carriage, seemed to make him envy me my son-in-law.

I was interested to see how Neil would handle Mossie. I was anxious that they would manage to find some accommodation because the farmer of Moss-side had been fundamental to the profits I had made in my last few years at Little Ardo. My successor handled my consultant to perfection. He waited until Mossie asked him, 'And how are you going to handle your crops, Neil?'

'I'm going to do exactly what you tell me.' He has done, and he has obviously done it better than I did, for he grew the four-tonne crop of barley to which I aspired for so long, in his first year and has continued to do so every year.

I remember my son-in-law's impatience with my preference for doing things in ways that were cheap but nasty. Like, I would work away with my loader, taking two bales at a time from the field to the barn rather than getting a squad together, a big cart and taking a load at a time. The new farmer of Little Ardo would want everything nice. He would have the best equipment. One of the first things he bought was a big second-hand tractor, a Ford 7740. It cost almost as much as my brand new Ursus, and he kept that as well. By the time the big Ford 7840 was home, Purdie had spent as much on machinery in two years as both the Allans had spent in fifty. And the big tractor was followed home by a much younger digger.

His grandfather-in-law John Allan would have been so pleased to see how Purdie really did do away with 'stupid labour' forty years after John had pronounced that it was all a thing of the past. In John's time, cattle in winter were bedded with straw trailed from the barn on a fork then carefully shaken out under the pairs of cattle tied in stalls. In my time we had cattle courts bedded with small round bales that had to be manhandled into the courts, and shaken out with forks. Then we got big bales. Two or three of those would be dumped by loader and the cattleman had to jump in amongst the beasts and roll out the bales before shaking them with the fork. He had to watch out, as the beasts boxed the bales as though they were invaders. As long as they weren't too wet and heavy that was a big advance. But now Purdie just loads a bale into his bedder, drives it to each pen and sits there while the great machine atomises the bale and blows it in, covering the pens in a shower of golden straw. Things cannot get any easier unless they get cows to drive tractors.

So that is all to the good and there is no doubt that Purdie has got Little Ardo back at last to where it had been in Maitland Mackie's time – right up to date with machinery. But the biggest changes have been wrought by the young man's building pro-gramme. He put up ten sheds in thirteen years. In this endeavour he did make use of one thing of mine, and that was my boyhood boozing pal Jock Paterson, the sma'hudder of Gowkstone just upstream from Little Ardo. A slater to trade, Jock became the master builder. Maitland Mackie's dairy byre got a new roof of boxed steel. And William Yull's barn got one too. The old byre – the one that was there when William Yull arrived in 1837, and whose roof was done when James Low arrived in 1930 – even it got a new roof. Then my Dutch barn, my joy when I built it in 1974 and my pride when James Low said I was now doing the things he had been at my father for years to do, got a lean-to at each side, which almost trebled its effective size. Then Maitland Mack-ie's great piggery, such a wonder in its time for its sheer length but far too low all my time in the farm, not only got the new, higher roof which I had contemplated, but a complete rebuild with portal frames, steel purlins, slatted sides for air vents . . . in fact, the only bit of the old shed that is left is part of the floor, and its name. It is still called the piggery, though now it holds cows and calves. My wonderful barley beef factory has had its slats removed and tanks filled to make cattle courts.

The table on the next page shows how the buildings at Little Ardo increased roughly tenfold from my great-great-grandfather's time. William Yull almost doubled the extent of the Little Ardo steading. His son John, the auctioneer, was less ambitious for the farm but liked his comforts. His great improvements were in the house.

But then, in 1911, came Maitland Mackie. He was the great builder. If he had been one to blaw he might have said he trebled the size of the place with his great piggery alone. But . . . there is a but: much of what Maitland Mackie built was not built to last. John Allan, as the table shows, was responsible for a decline in the size of the steadings, mainly the two large black creosoted henhouses. They would have been an eyesore anywhere, but two things increased the problem. They were so close to the house that one of them almost came in the back door. It was a great improvement when John pulled that down in 1946. But the other one had a special fault. I have told you that Maitland Mackie was careful with his money. Well, when they built the black sheddies, James Low ran out of creosote when he had still six feet to preserve. Maitland Mackie told his grieve that there was no money for more creosote. So by 1945 that end was beginning to deteriorate. By 1960 the eyesore had become a source of ridicule and, mercifully, it was also taken down. Maitland's wonderful piggery was also wooden, and despite John Allan giving it breeze-block walls it was really finished when Neil Purdie replaced it in 2003. I was so sorry that my mother hadn't lived to see the day when Maitland's new barn finally blew down in 2005, to no one's regret except the insurance company's.

My turn of the overdraft led to a 40 per cent increase in buildings, and then came the second great builder, and one who built to last. On the face of it, Purdie's 60 per cent increase in the steadings may not compare with Maitland Mackie's – but look again. He has also replaced the piggery, and the old byre, and re-roofed the new byre and the old barn as well as adding his 60 per cent of new build.

Little Ardo: farmers and extent

Farmer	Date started	Final acreage	Steading (sq feet)	Percentage change
Unknown	?	150	3,983	
William Yull	1837	150	7,793	+96%
John Yull	1869	150	7,793	+0%
Dod Yull	1905	150	7,793	+0%
Maitland Mackie	1911	232 +55%	21,130	+171%
John Allan	1945	232	18,062	−15%
Charlie Allan	1973	252 +9%	25,296	+40%
Neil Purdie	1997	252	40,473	+60% so far

You will see an imbalance in the progress of the little farm on the hill. While it now has a very large modern steading, up tenfold in our family's time here, the acreage has only increased by 67 per cent.

Another point about Little Ardo is that it was always supported by cash flows from off the farm rather than supporting the farmers in the style of gentlefolk. William Yull, the first of us, may even have been the only one of our family of farmers who was dependent on Little Ardo rather than the other way round. John Yull, his son, was a very successful auctioneer and valuer. Dod Yull was perhaps a dedicated farmer, except that what he was truly dedicated to was hot-footing it to the New World as soon as possible. Then Little Ardo became part of Maitland Mackie's empire, and while that meant it was dedicated to producing food and profit if possible, the farmer was by no means dependent on it – he had plenty of other farms. When John Allan came home from the war he had a few profitable years on the farm. He may even have made more from farming in some of the boom years after the war than he made from his writing and broadcasting, but that didn't last long. Then there was me. I made money off the cattle when I was farming by telephone and paying the bills with a senior lecturer's pay from Strathclyde University. But by the time I came home to live at Backhill and farm by the sweat of my brow, the boom in fancy cattle was over. My wife saw immediately that this was not going to work and finished her qualification as a teacher.

A couple of years later and I was singing, writing for the papers, publishing books and *Leopard* magazine and I even took a full-time job at the BBC. I was working like mad to keep the farm, not the other way round. And finally we have Neil Purdie. His wife has a chemist's shop . . .

The traditional relationship between the old farmer and his successor is of the old man shaking his head at the folly he surveys. A man with a sense of history like me should descend on the farm at regular intervals and sweep through the byres and the sheds saying, 'That's wrang. That's wrang. And that's awfu wrang.' To conform to what is expected of me I should now be revelling in the mistakes Young Purdie is making. I should be taking revenge on the next generation for all the head-shaking Jimmy Low did in my time. It was Andrew Dunlop, the Ayrshire dairyman, who suggested this relationship should be acknowledged in the name of our new house which should, he suggested, be Glowerindoon, rather than Dunfermin or, as we had pencilled in, Whinhill. Neil Purdie also had a suggestion; 'Middenview'. I stupidly pointed out that though we could see the steading all right we couldn't really see the midden. 'Not yet,' he said, with more than a hint of menace.

After thirteen years of glowering down on the family farm there is still no midden over the dyke from Whinhill. I take no credit, because the new farmer has hardly done anything of which I disapprove. You might say that Neil likes everything so tidy that there isn't enough left for the wildlife, but that isn't a serious issue with me. I know that many people's idea of beauty in a landscape is wilderness. I have often seen what they mean. I have seen it in the snows of Alaska, the deserts of North Africa, the Italian Alps and even in the west coast of Scotland. But to the farmer of Scotland's North-east, beauty in the land is order. It is the triumph of his choice of fauna and flora over that which would have occurred in his absence. So while wide field margins are good for the partridges and unkempt road verges are good for the voles on which the barn owls swoop so eagerly, my farmer's eye does enjoy the way Little Ardo looks now. And who could fail to relish the fact that there is not only a close brush but one that is used to brush the close?

The Purdies have really accomplished all but one of the things towards which the Yulls, the Mackies and the Allans strove at Little Ardo. It has good stock, it has smashing machinery, good handling pens, enough accommodation for a medium-sized airport and concrete 'roadies aa roon and roon'. The Single Hoose to which Maitland Mackie brought running water and John Allan brought electricity, is right at the top end of the holiday lets market, with central heating and a treble-glazed sitooterie commanding views up and down the river, over the village and across Formartine to Bennachie.

There is just one want left about the place. It is still too small to give the farmer the chance of a good living. But then, the Purdies aren't finished yet.

So here we are, ideally placed in the neuk of the park to watch the light fading slowly over the family farm and my people's country.

I hope it only sounds a little smug, but I am happy that I have fulfilled the peasant's first duties. I have seen Little Ardo into safe and willing hands. And I have seen to it as far as I can, that the family will always have this place to call their place. If those 'duties' are bunk it is John Allan's fault. In his novel, *Green Heritage*, he put this in the mouth of Stephen, his hero:

> No man can love a whole wide country and the many millions that live there. Such patriotism is a vague and sentimental thing. The true patriot is the man whose roots are in some small corner of the earth where his fathers have added beauty and enriched it with their work. The man who has some such place to love is like a tree whose roots are deep in native earth; the man without is driftwood blown about the world.

As women live longer than men, the last word is likely to be with Fiona, so the last word here should be about her. She made it possible for me to follow my philosophy that No Job Should Be A Life Sentence. She enabled me to give up what seemed like promising careers to follow others even if they might butter no beans. And she showered me with so many gifts: four children who were never boring and always loving; a sports car for my twenty-eighth birthday; a muck spreader for my thirty-eighth and now, in my seventy-first year, she tells me she is saving for an electric buggie so that I can drive myself along the pavement to the pub or to watch the cricket.

Index